Springer Proceedings in Materials

Volume 47

Springer Proceedings in Materials publishes the latest research in Materials Science and Engineering presented at high standard academic conferences and scientific meetings. It provides a platform for researchers, professionals and students to present their scientific findings and stay up-to-date with the development in Materials Science and Engineering. The scope is multidisciplinary and ranges from fundamental to applied research, including, but not limited to:

- Structural Materials
- Metallic Materials
- Magnetic, Optical and Electronic Materials
- Ceramics, Glass, Composites, Natural Materials
- Biomaterials
- Nanotechnology
- Characterization and Evaluation of Materials
- Energy Materials
- Materials Processing

To submit a proposal or request further information, please contact one of the following Springer Publishing Editors according to your affiliation location:

European countries: **Mayra Castro** (mayra.castro@springer.com)
India, South Asia and Middle East: **Priya Vyas** (priya.vyas@springer.com)
South Korea: **Smith Chae** (smith.chae@springer.com)
Southeast Asia, Australia and New Zealand: **Ramesh Nath Premnath** (ramesh.premnath@springer.com)
The Americas: **Michael Luby** (michael.luby@springer.com)
China and all the other countries or regions, as well as topics in materials chemistry: **Maggie Guo** (maggie.guo@cn.springernature.com)

This book series is indexed in **SCOPUS** and EI Compendex database.

S. T. Aruna · Ashwini Ravi · U. Kamachi Mudali
Editors

Recent Advances in Electrochemical Science and Technology

Select Proceedings of 1st International Conference on Women in Electrochemistry (ICWEC) 2023

Springer

Editors
S. T. Aruna
Surface Engineering Division
CSIR-National Aerospace Laboratories
Bengaluru, Karnataka, India

Ashwini Ravi
Medetronix Labs Pvt. Ltd. (MTX)
Bengaluru, Karnataka, India

U. Kamachi Mudali
Homi Bhabha National Institute (HBNI)
Mumbai, Maharashtra, India

ISSN 2662-3161 ISSN 2662-317X (electronic)
Springer Proceedings in Materials
ISBN 978-981-97-3957-8 ISBN 978-981-97-3958-5 (eBook)
https://doi.org/10.1007/978-981-97-3958-5

This Springer imprint is published by the registered company Springer Nature Singapore Pte Ltd.
The registered company address is: 152 Beach Road, #21-01/04 Gateway East, Singapore 189721, Singapore

Contents

About the Editors

Dr. S. T. Aruna is Chief Scientist at Council of Scientific and Industrial Research—National Aerospace Laboratories (CSIR-NAL), Bengaluru, India. She completed her master's degree in chemistry from Mysore University (1994), where she secured the first rank and won four gold medals. She obtained her Ph.D. from the Indian Institute of Science (IISc), Bengaluru in 1998. She worked as a postdoctoral fellow at Bar-Ilan University, Israel in dye-sensitized solar cells from 1998 to 1999. She has authored ~130 research papers in international peer-reviewed journals, filed 11 patents, co-authored a book and coedited a book, written 14 book chapters and four encyclopedia chapters. Her research papers have been well-cited (10,255), and her publications have an h-index of 44. Her name features in the list of the top 2% of world scientists released by Stanford University. Her current research interests include the development of tapecast ceramic substrates for electronics application, nanomaterials, plasma-sprayed thermal and environmental barrier coatings, and oxygen sensor and solid oxide fuel cells (SOFCs). She has guided five Ph.D. and 25 master's and B.Tech. students. One of her review papers has received the best-cited paper award twice, which was published in *Current Opinion in Solid State and Materials Science*. She is on the editorial board of *Surface Engineering Journal*. She has won Sathish Dhawan Young Engineer Karnataka State Award 2021, Sasadhar Ray Memorial award 2023, MRSI Medal 2022, NM Sampat award 2016, and Pavan Nagpal Memorial award 2019, among many others.

Dr. Ashwini Ravi is Quality and Application Analyst at Medetronix Labs Pvt. Ltd. She obtained her Ph.D. in Nanotechnology from Visvesvaraya Technological University (VTU), Belgaum, Karnataka. She holds master's and bachelor's in nanotechnology and biotechnology, respectively, both from VTU. Her passion for scientific research drove her to the Department of Materials Engineering at IISc as Research Assistant to work on 2D nanomaterials, such as graphene and its derivatives for biomedical applications. Later, she moved to the Centre for Biosystems, Science, and Engineering (BSSE), IISc and worked as Research Assistant for couple of years on bioimaging techniques and stem cells-based 3D nanofibrous scaffolds. Her research experience at various departments in IISc helped her generate research collaborations and excel in her research career. These experiences provided an impetus to begin doctoral research work on the development of graphene and its derivatives-based electrochemical sensors. She was able to publish her research work in some international journals of high repute, attracting significant interest among the research community as reflected by the citations received. Additionally, her research work was recognized by Department of Science and Technology (DST), Goverment of India, which sanctioned her project under the Woman Scientist Scheme-B (WOS-B) for 3 years. Moreover, she was awarded the prestigious "Raman Charpak Fellowship" under Indo–French bilateral Fellowship Programme to work on her doctoral research at IEMN labs, University of Lille, France. She also actively takes part in organizing scientific workshops, international and national conferences, symposiums, and webinars in association with non-profit science organizations.

Dr. U. Kamachi Mudali Vice Chancellor of Homi Bhabha National Institute, DAE is formerly Vice Chancellor of VIT Bhopal and Distinguished Scientist of Department of Atomic Energy (DAE), and Chief Executive and Chairman of Heavy Water Board, a flagship industrial unit of DAE. Dr. Mudali has made pioneering contributions to advanced materials and coating technology; corrosion science, engineering, and technology; materials, process, and equipment development for reprocessing applications; failure analysis, consultancy, and societal contributions, for nearly four decades during his service, including as Director of Materials Chemistry and Metal Fuel Cycle at Indira Gandhi Centre for Atomic Research, DAE, Kalpakkam. Dr. Mudali has 474 journal publications, five patents, and 13,125 citations with h-index of 55 and i-10 index of 308. He is recognized in the World's Top 2% scientists from India in the field of materials and is a fellow of 12 professional associations, including four from abroad. Dr. Mudali is decorated with several distinguished recognitions from India and abroad including: Distinguished Alumnus Award from IIT Bombay, where he did M.Tech. in corrosion science and engineering in 1984; PSG College of Technology, Coimbatore, where he did M.Sc. in materials science in 1982; Lifetime Achievement Award from Ministry of Steel, Government of India, Electrochemical Society of India, Bengaluru and National Corrosion Council of India, Karaikudi; GD Birla Gold Medal, Platinum Medal and Metallurgists of the Year Award from IIM and Ministry of Steel, MASCOT National Award, Vasvik Award, INS Medal, Homi Bhabha Science and Technology Award of DAE; Frank Newman Speller Award of NACE International, USA, the highest recognition and first Indian to get in the field of corrosion engineering.

Solution Combustion Synthesized and Tape Cast $(Sc_{0.1}Ce_{0.01}Zr_{0.89}O_{2-\delta})_{0.98}(Al_2O_3)_{0.02}$ Electrolyte for SOFC Application

J. Ramya, R. Packiaraj, A. Sreelakshmi, S. Senthil Kumar, and S. T. Aruna

1 Introduction

The exponential growth of renewable energy demand has enabled solid oxide fuel cells (SOFCs) to reach the commercial scale [1]. SOFC is regarded as one of the most appealing energy conversion devices for converting chemical energy into electrical energy with high efficiency. Zirconia is a widely preferred electrolyte for the SOFC due to its long-term stability at high temperatures. The stabilized zirconia (ZrO_2) has high oxygen ion conductivity over a wide range of temperatures and oxygen partial pressures, making it a suitable material for SOFC electrolytes. 8 mol% yttria-stabilized zirconia (YSZ) has been playing a dominant role as electrolyte over many years and is a promising electrolyte for anode supported SOFC [2–6]. However, in order to realize reasonable performance in an electrolyte-supported cell (ESC), YSZ has to be operated in the temperature range of 900–1000 °C. Nevertheless, the long-term exposure to such high temperatures causes overall degradation of the stack components [7–9]. So there is a need to find alternative electrolytes with better ionic conductivity to reduce the operating temperatures to below 800 °C and the fabrication process is easily adapted for its commercial-scale production.

In a typical operating condition of the SOFC, the major part of ohmic resistance arises from the electrolyte material. Usually, compounds with high crystal site symmetry are preferred for good ionic conduction. The zirconia can be stabilized in a high-symmetry cubic phase by using suitable dopants. The concentration of the dopant dictates the ionic conductivity of ZrO_2-based electrolytes. It is well

J. Ramya · R. Packiaraj · A. Sreelakshmi · S. Senthil Kumar (✉) · S. T. Aruna (✉)
Surface Engineering Division, Council of Scientific and Industrial Research-National Aerospace Laboratories, Post Bag No. 1779, Bangalore 560017, India
e-mail: ssenthil@nal.res.in

S. T. Aruna
e-mail: aruna_reddy@nal.res.in

S. T. Aruna et al. (eds.), *Recent Advances in Electrochemical Science and Technology*, Springer Proceedings in Materials 47, https://doi.org/10.1007/978-981-97-3958-5_1

known that the maximum ionic conductivity is commonly realized near to minimum required concentrations of dopants that are required to completely stabilize the cubic fluorite structure. However, very high levels of dopant concentrations can induce the ordering of vacancies and increase the association of dopants and oxygen vacancies into complex defects with low mobility. In general, rare-earth ion doping is more effective than alkaline earth elements doping [1].

The ionic conductivity of doped zirconia reduces with increasing the ionic radii of rare-earth ion dopants [2]. The ionic radius of Sc^{3+} is lower than other rare-earth elements and close to Zr^{4+}. Hence, scandia (Sc_2O_3) stabilized zirconia (ScSZ) yields better ionic conductivity than the state-of-the-art yttria-stabilized zirconia (YSZ) [10–12]. Scandia dopant in the range of 7–11 mol% exhibits the secondary rhombohedral phase which is a lower ionic conducting phase. Further, this rhombohedral phase undergoes reversible transformation to the cubic phase above 550 °C. This transformation induces cracks in the electrolyte and affects the structural integrity of the cells. To overcome this, the co-doping of ceria has been adapted to control transformation and retain the highly symmetric cubic structure [3].

Further, achieving the required mechanical strength is equally challenging in the thinner (<200 μm) electrolyte-supported cells [4]. There are reports on improving the mechanical properties of ScSZ electrolyte using sintering aids like CeO_2, Gd_2O_3, Sm_2O_3, Bi_2O_3, Yb_2O_3, and Al_2O_3 [13–15]. It is also evident from the literature that the addition of a small amount of alumina ($\leq$2 wt%) can promote the sintering effectively. Lakshmi et al. [7] studied the effect of Yb_2O_3 co-doping in ScSZ during solution combustion synthesis on phase formation and electrical conductivity. The co-doped Yb suppressed the secondary phase and improved the ionic conductivity. Grosso and Muccillo [5] demonstrated the stabilization of the cubic phase with 1% Ce doping. High temperature and long-duration sintering also eliminated the secondary phase. Sriubas et al. [6] prepared the scandia and alumina co-doped zirconia having the secondary phase by electron beam physical vapor deposition (EBPVD), which exhibited an electrical conductivity of 4.2×10^{-3} S/cm. Therefore, from the literature, it is evident that co-doping of ScSZ with appropriate dopant concentrations is the best way to realize the electrolyte with higher ionic conductivity and mechanical strength suitable for operation at 650–800 °C. Also, it is evident that there is no literature on solution combustion synthesis of co-doped ScSZ powder suitable for tape casting, cell fabrication, and performance evaluation.

In the present work, Sc and Ce co-doped zirconia containing alumina (10ScCeSZ-Al) as the sintering aid was synthesized by the versatile single-pot solution combustion method to attain better homogeneity and phase purity. Further, an attempt was made to utilize the synthesized powder in the industrially scalable process like tape casting and preparing thin tapes followed by cell fabrication and evaluating its electrochemical properties.

2 Experimental

2.1 Synthesis and Characterization of 10ScCeSZ-Al

Scandia, ceria, and alumina co-doped zirconia (10ScCeSZ-Al) were chosen as the electrolyte's composition. The electrolyte powder was prepared through the solution combustion synthesis of nitrate salts using citric acid as the fuel. Nitrate and chloride salts were used as precursors because they are easily soluble in water and react to form metal nitrate compounds at a few hundred degrees Celsius [16] while liberating chlorine and nitrogen gases along with water vapor during a typical combustion reaction. The procedure is outlined in Fig. 1. For the synthesis $ZrOCl_2{\cdot}8H_2O$ (Loba Chemie) and Sc_2O_3 (Chemsworth) were used as the starting materials. Separately, deionized water was used to dissolve $ZrOCl_2{\cdot}8H_2O$, $Al(NO_3)_3{\cdot}9H_2O$, $(NH_4)_2Ce(NO_3)_6$, and $C_6H_8O_7{\cdot}H_2O$ in a 1 L glass beaker while concentrated nitric acid was used to dissolve Sc_2O_3. The oxidizer to citric acid fuel ratio was maintained as unity [17]. Gradually, each solution was added one by one at a time into the zirconium oxychloride solution, heated to reduce the volume of the solution on a hot plate, and then transferred to a muffle furnace that was preheated to 450 °C. The solution underwent a smoldering type of combustion to give a powder containing traces of carbon. Hence the powder was subjected to calcination at 900 °C for 4 h (at a ramp of 3 °C/min). The calcined powder was wet milled to reduce the particle size using a Fritsch planetary ball mill with distilled water as a solvent and zirconia balls as milling media. The milling was done for 5 h at 350 rpm. Henceforth, the prepared co-doped zirconia electrolyte composition and commercial Tosoh 8 mol% yttria-stabilized zirconia are referred to as 10ScCeZr-Al and 8YSZ, respectively. The phase purity of the powder was analyzed by X-ray diffractometry using Bruker D-8 Advance X-ray diffractometer. The average crystallite size of ScSZ was calculated using the Debye–Scherrer formula $D_{hkl} = K\lambda/(\beta_{hkl} \cos \theta)$ where D_{hkl} is the crystallite size in the direction perpendicular to the lattice planes, hkl are the Miller indices of the planes being analyzed, K is a numerical factor frequently referred to as the crystallite-shape factor, λ is the wavelength of the X-rays, β_{hkl} is the width (full-width at half-maximum) of the X-ray diffraction peak in radians and θ is the Bragg angle. The particle size distribution of ScSZ powder was measured using a particle size analyzer (Malvern Mastersizer 2000). A Carl Zeiss field emission scanning electron microscope (FESEM) was used to obtain the morphology and microstructure of the synthesized powder and sintered tape, respectively. The relative density of the sintered pellet was measured using Archimedes' method.

2.2 Button Cell Fabrication and Testing

The ball-milled powder was used to fabricate 10ScCeSZ-Al tape for SOFC fabrication. The slurry for tape casting was prepared using the ball-milled fine powder.

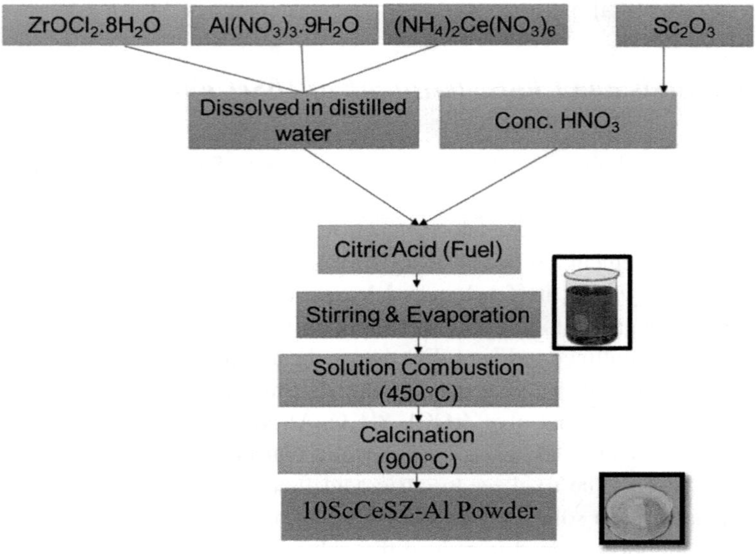

Fig. 1 Flowchart depicting the steps involved in the synthesis of 10ScCeSZ-Al powder by solution combustion method

Suitable binders and precursors were added to the slurry and milled using a roller mill as reported elsewhere [18]. The slurry was cast on a Mylar sheet to a thickness of 220 microns using a doctor's blade. The tape-casting process takes place by feeding the slurry through the doctor's blade, which screeds the slurry into a thin layer of fixed thickness, allowing its rapid drying by evaporation. After drying, the tape was cut into a circular shape with a diameter of 30 mm. Then 20 tapes were pressed at a pressure of 90 MPa in a suitable die. The obtained pellet was sintered at 1525 °C for 2 h (3 °C/min). Then the anode paste (62.5 wt% NiO-37.5 wt% 8YSZ, in-house prepared composition) was printed on one side and was fired at 1250 °C/2 h (at a rate of 3 °C/min). Cathode paste (50 wt% $L_{0.65}S_{0.35}MnO_{3-\delta}$-50 wt% ScSZ, in-house prepared composition) was printed on the other side of the pellet and sintered at 1100 °C/2 h (at a rate of 3 °C/min). The button cells were fabricated according to the schematic of the cell layers shown in Fig. 2. The electrical conductivity of the synthesized 10ScCeSZ-Al electrolyte was evaluated and compared with the commercial Tosoh 8YSZ electrolyte processed under similar conditions. The electrical conductivity which is regarded to be the first priority among the parameters that an electrolyte should meet was measured using Vander Pauw 4 probe method using the Keithley current source and nano-voltmeter at various temperatures between 500 and 700 °C [8].

For electrochemical characterizations, the button cells with 1 cm^2 active area were placed on a test bench using ceramabond 552 sealant. The sample in the SOFC test bench (FCT, USA) was subjected to 850 °C (2 °C/min) in the air followed by a 4 h reduction in 20: 80 H_2–N_2, and the cell performance was studied in dry H_2 on 1

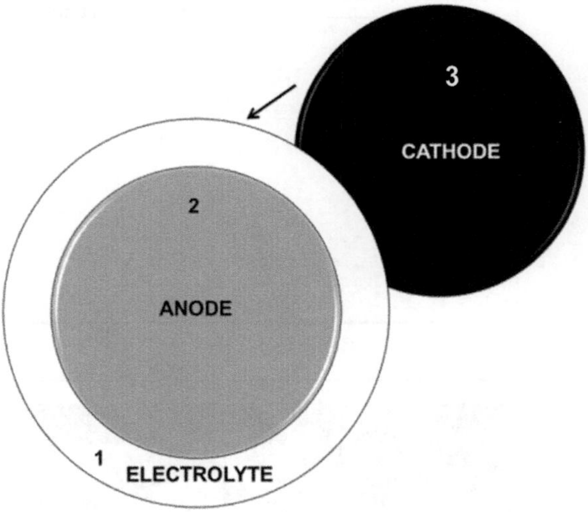

Fig. 2 Schematic depicting the button cell components and the numbers indicate the order of fabrication

cm^2 button cell. The fuel and the O_2 flow rates were maintained at 50 and 100 sccm, respectively. The cells could not be recovered after the cell testing because of the sealant. For every test, 2–3 cells were tested in similar conditions, and performance was tested in H_2 at 800 and 850 °C.

Impedance studies were carried out by using a CH electrochemical analyzer (CHI 604 2D) controlled by CHI604D software for impedance measurements of the button cell. The experiment was conducted under the OCV condition and impedance studies were conducted in 1 MHz to 0.1 Hz frequency range with a voltage of 10 mV amplitude.

3 Results and Discussions

3.1 Microstructure, Morphology, and Structure Characterization

The fine powder was obtained using the ball milling process. Particle size was measured using well-dispersed suspension. An average agglomerated particle size of d_{50} of 1.53 μm was achieved for the ball-milled powder and it showed a unimodal distribution with d_{10} and d_{90} of 0.59 μm and 4.16 μm, respectively (Fig. 3a). The FESEM image of ball-milled 10ScCeSZ-Al powder shows irregularly shaped particles in the size range of 0.5–1.5 μm (Fig. 3b), in agreement with the particle size analysis results (Fig. 3a).

Figure 4 shows the XRD patterns of synthesized 10ScCeSZ-Al powder and sintered pellet. The XRD patterns of powder and sintered pellets match well with the cubic ScSZ pattern of the ICSD database (JCPDS card no. 01-089-5485). No traces

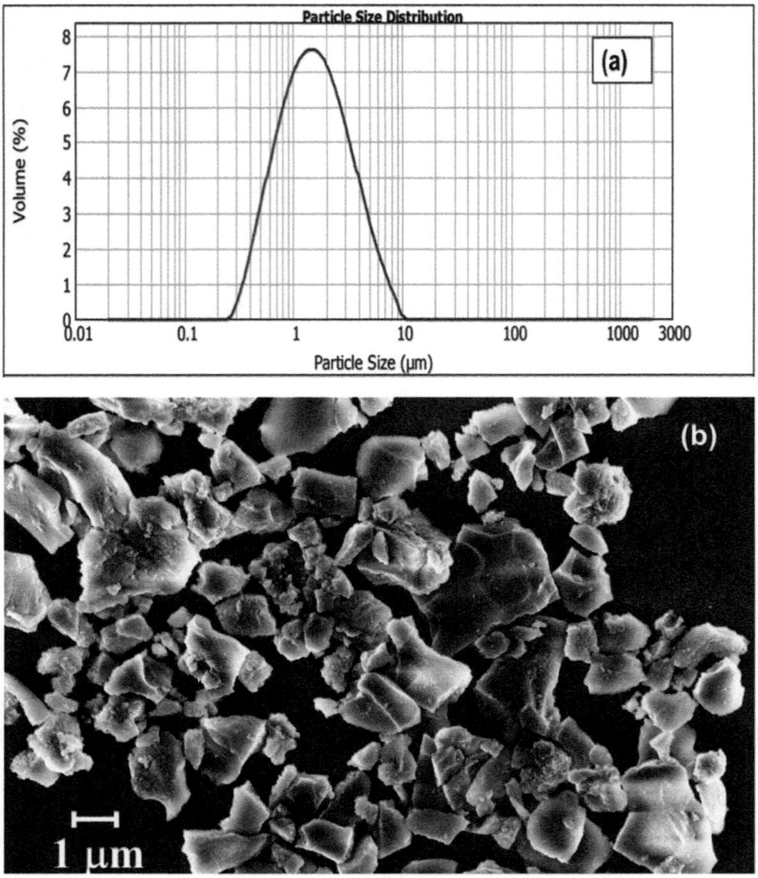

Fig. 3 **a** Particle size distribution of 10ScCeSZ-Al powder and **b** FESEM image of ball-milled 10ScCeSZ-Al powder

of reactant peaks such as alumina, ceria, etc., were detected, indicating the formation of co-doped zirconia. The as-synthesized powder had broad peaks. However, there was no peak shift observed in 10ScCeSZ-Al powder with respect to the standard ScSZ phase (JCPDS: 01-089-5485). A very small shift of 2θ~0.2 occurred in the sintered 10ScCeSZ-Al pellet which may be attributed to the slight change in lattice parameters. The formation of cubic phase is identified from the peak profile in the angular region between 73° and 75° [5]. The crystallite size of ball-milled 10ScCeSZ-Al powder calculated from the Debye–Scherrer formula was 15 nm.

The FESEM image of the sintered 10ScCeSZ-Al pellet is shown in Fig. 5. The sintered pellet had dense close-packed grains with distinct grain boundaries. The grain size was in the range of 2–5 μm. The pellet appeared to be well-sintered with less porosity. Further increase in sintering temperature or time can eliminate the porosity completely.

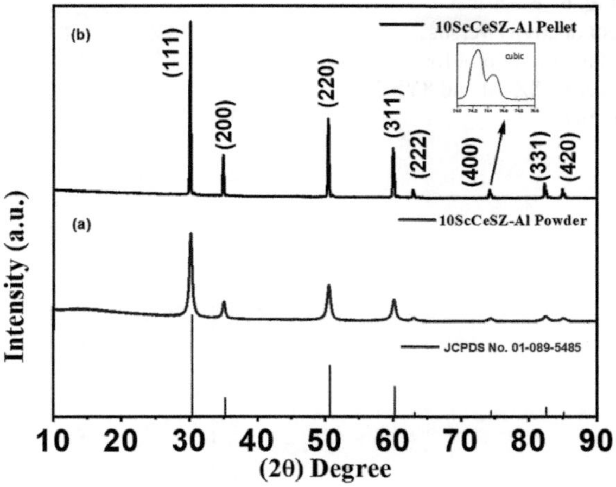

Fig. 4 XRD patterns of 10ScCeSZ-Al **a** calcined powder and **b** pellet sintered at 1525 °C

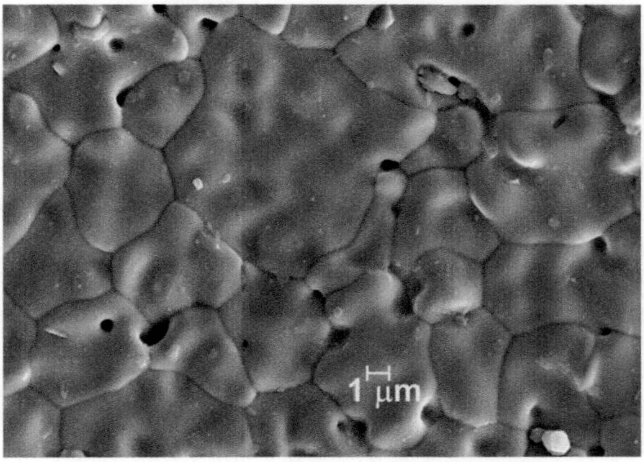

Fig. 5 FESEM image of sintered 10ScCeSZ-Al pellet

3.2 *Electrical conductivity of Sintered 10ScCeSZ-Al Electrolyte*

Figure 6 shows the electrical conductivity versus temperature plot for the 10ScCeSZ-Al. The 10ScCeSZ-Al pellet showed an electrical conductivity of 0.029 S/cm at 700 °C whereas for 8YSZ electrical conductivity is 0.016 S/cm. Grosso and Muccillo [5] reported the conductivity of 10ScCeSZ without alumina doping, which is almost half (0.015 S/cm) of the present composition. It is clearly evident that alumina doping has increased the electrical conductivity, which is critical to cell performance. The activation energy has been derived by fitting the experimental results based on the following Arrhenius function.

$$Ln(\sigma) = Ln(\sigma_o) + (E_a/R)(1/T) \tag{1}$$

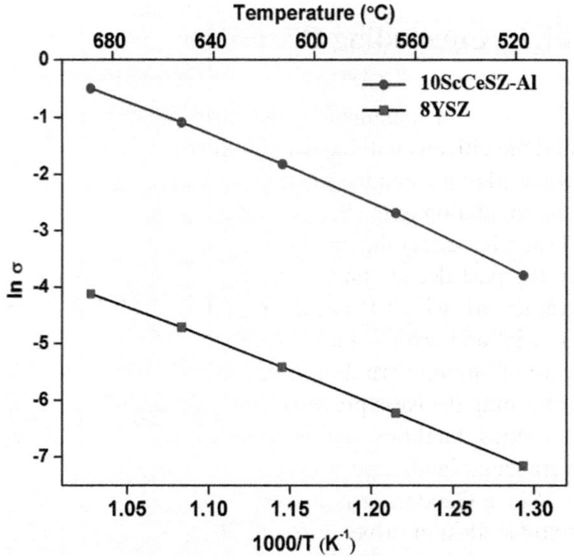

Fig. 6 Electrical conductivity versus temperature plot for 10ScCeSZ-Al and 8YSZ pellets

where E_a is the activation energy, R is the universal gas constant, and T is the absolute temperature.

The activation energy (E_a) for the 10ScCeSZ-Al is 1.06 eV, which was close to 8YSZ (0.986 eV) and similar to the reported value in the literature [19, 20]. Though the activation energy is the same, the higher pre-exponential factor (σ_o) of Eq. 1 resulted in high electrical conductivity for 10ScCeSZ-Al.

3.3 *Button Cell Testing*

The performance of the button cells with both commercial 8YSZ and synthesized 10ScCeSZ-Al electrolytes tested under similar conditions is presented in Fig. 7. Both the 8YSZ and 10ScCeSZ-Al cells exhibit a high OCV of 1.2 V. From the plot it is evident that the 10ScCeSZ-Al based SOFC shows a power density of 191 mW/cm^2 at 850 °C, 155 mW/cm^2 at 800 °C, and 133 mW/cm^2 at 775 °C. YSZ-based SOFC shows a power density of 88 mW/cm^2 at 800 °C. The improved performance of 10ScCeSZ-Al can be attributed to the higher ionic conductivity of 10ScCeSZ-Al. Due to the higher conductivity of the 10ScCeSZ-Al, there was a reduction in the ohmic resistance of the cell. In addition, the use of 10ScCeSZ-Al in electrodes has reduced the polarization resistance. There was about a 76% reduction in the cell‘s Area Specific Resistance (ASR) which has improved the overall cell performance by about 76%.

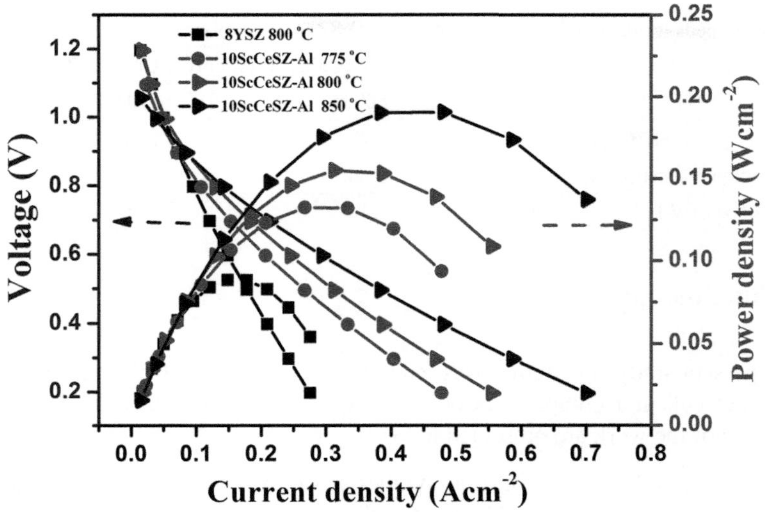

Fig. 7 SOFC button cell performance with 10ScCeSZ-Al as the electrolyte support

The impedance plots are shown in Fig. 8. From the plot, it is evident that the 10ScCeSZ-Al button cell shows higher impedance compared to the 8YSZ-based SOFC tested under similar conditions. The impedance data are tabulated in Table 1.

Fig. 8 Impedance spectra for 10ScCeSZ-Al and 8YSZ-based SOFCs

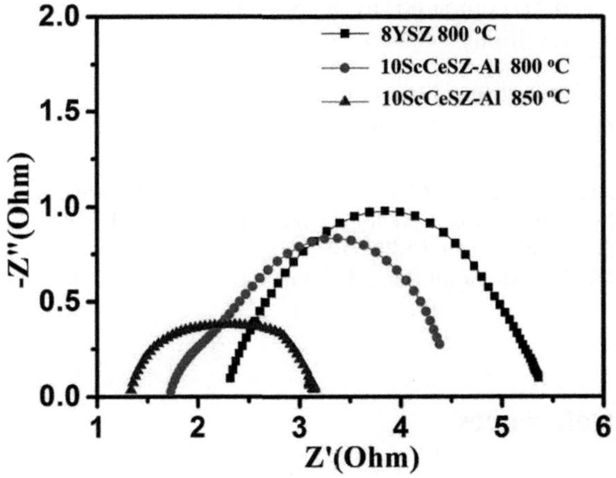

Table 1 Impedance data of Tosoh 8YSZ and 10ScCeSZ-Al electrolytes

Resistance values	8YSZ (Ω cm)	10ScCeSZ-Al (Ω cm)	10ScCeSZ-Al (Ω cm)
	800 °C	800 °C	850 °C
Ohmic resistance (R_{Ohm})	2.31	1.72	1.31
Polarization resistance (Rp)	3.05	2.66	1.82
ASR (slope of V-I plot)	5.36	1.42	1.05

4 Conclusions

In the present study, an attempt was made to fabricate a 10ScCeSZ-Al-based solid oxide fuel cell and compare its performance vis-à-vis 8YSZ-based SOFC. The highlights of the work are presented below:

- Single-step solution combustion synthesis was successful for the synthesis of single-phase cubic Ce and Al-doped ScSZ.
- The synthesized powder was agglomerated and the particle size was reduced by ball milling and the powder after ball milling exhibited an average agglomerated particle size of 1.534 μm and was suitable for tape casting.
- The electrical conductivity of the 10ScCeSZ-Al electrolyte was found to be 0.029 S/cm which was higher than the commercial 8YSZ (0.016 S/cm) processed and tested under the same conditions.
- The 10ScCeSZ-Al-based button cell exhibited higher power density (191 mW/cm^2) compared to 8YSZ-based SOFC (88 mW/cm^2) tested under similar conditions.

Thus, the solution combustion process is a viable technique for the synthesis of tape-cast grade Ce and Al-doped ScSZ powder suitable for SOFC application.

Acknowledgements The authors acknowledge the grants received from DST under the Advanced hydrogen fuel cell program (Grant no.: DST/TMD-EWO/AHFC-2021/2021/118). The authors thank the Director, CSIR-NAL, and Head, SED for the encouragement. The authors are grateful to Mr. Siju, Mr. Srinivas and Mr. Balaji, for FESEM, XRD and particle size analysis respectively.

References

1. Choolaei M, Vostakola MF, Horri BA (2023) Recent advances and challenges in thin-film fabrication techniques for low-temperature solid oxide fuel cells. Crystals 13:1008. https://doi.org/10.3390/cryst13071008
2. Prerna V, Khandla M, Chaudhary K, Pati R (2023) Recent advances on electrolyte materials for SOFC: a review. Inorg Chem Comm 152:110724. https://doi.org/10.1016/j.inoche.2023.110724
3. Zakaria Z, Abu Hassan SH, Shaari N, Yahaya AZ, Kar YB (2020) A review on recent status and challenges of yttria stabilized zirconia modification to lowering the temperature of solid oxide fuel cells operation. Int J Energy Res 44(2):631–650. https://doi.org/10.1002/er.4944

4. Guo CX, Wang JX, He CR, Wangn WG (2013) Effect of alumina on the properties of ceria and scandia co-doped zirconia for electrolyte-supported SOFC. Ceram Int 39:9575–9582. https://doi.org/10.1016/j.ceramint.2013.05.076
5. Grosso RL, Muccillo ENS (2013) Sintering, phase composition and ionic conductivity of zirconia-scandia-ceria. J Power Sourc 233:6–13. https://doi.org/10.1016/j.jpowsour.2013.01.140
6. Sriubas M, Kainbayev N, Virbukas D, Bockute K, Rutkuniene Ž, Laukaitis G (2019) Structure and conductivity studies of scandia and alumina doped zirconia thin films. Coatings 9:317. https://doi.org/10.3390/coatings9050317
7. Vlasov AN, Perfilliev MV (1987) Phase formation and ionic conductivity studies on ytterbia co-doped scandia stabilized zirconia ($0.9ZrO_2$–$0.09Sc_2O_3$–$0.01Yb_2O_3$) electrolyte for SOFCs. Solid State Ion 13(8):1520–1525. https://doi.org/10.1016/j.solidstatesciences.2011.05.014
8. Hengyong T, Stimming U (2004) Advances, aging mechanism and lifetime in solid-oxide fuel cells. J Power Sources 127:284–293. https://doi.org/10.1016/j.jpowsour.2003.09.025
9. Singh RN, Chavan SV (2007) Processing and properties of scandia-doped zirconia electrolyte for intermediate temperature SOFC. ECS Trans 7(1):2207–2212. https://doi.org/10.1149/1.2729336
10. Jais AA, Ali SAM, Anwar M, Somalu MR, Muchtar A, Isahak WNRW, Baharudin NA, Lim KL, Brandon NP (2020) Performance of Ni/10Sc1CeSZ anode synthesized by glycine nitrate process assisted by microwave heating in a solid oxide fuel cell fueled with hydrogen or methane. J Solid State Electrochem 24(3):711–722. https://doi.org/10.1007/s10008-020-04512-6
11. Steele BCH (1989) Oxygen ion conductors. In: Takahashi T (ed) High conductivity solid ionic conductors: recent trends and applications. World Scientific, Singapore, pp 402–446
12. Etsell TH, FIengas SN (1970) Electrical properties of solid oxide electrolytes. Chem Rev 70:339–376. https://doi.org/10.1021/cr60265a003
13. Sarat S, Sammes N, Smirnova A (2006) Bismuth oxide doped scandia-stabilized zirconia electrolyte for the intermediate temperature solid oxide fuel cells. J Power Sources 160:892–896. https://doi.org/10.1016/j.jpowsour.2006.02.007
14. Ishii T (1995) Structural phase transition and ionic conductivity in $0.88ZrO_2$ $(0.12-x)Sc_2O_3xAl_2O_3$. Solid State Ion 78:333–338. https://doi.org/10.3390/coatings10090870
15. Wang Z, Cheng M, Bi Z, Dong Y, Zhang H, Zhang J, Feng Z, Li C (2005) Structure and impedance of ZrO_2 doped with Sc_2O_3 and CeO_2. Mater Lett 59:2579–2582. https://doi.org/10.1016/j.matlet.2004.07.065
16. Vijaya Lakshmi V, Bauri R, Gandhi AS, Paul S (2011) Synthesis and characterization of nanocrystalline ScSZ electrolyte for SOFCs. Int J Hydrog Energy 36(22):14936–14942. https://doi.org/10.1016/j.ijhydene.2011.02.139
17. Aruna ST, Mukasyan AS (2008) Combustion synthesis and nanomaterials. Curr Opin Solid State Mater Sci 12:44–50. https://doi.org/10.1016/j.cossms.2008.12.002
18. Senthil Kumar S (2016) Studies on internal reforming solid oxide fuel cell. PhD thesis, Indian Institute of Science, India
19. Zhang J, Lenser C, Menzler NH, Guillon O (2020) Comparison of solid oxide fuel cell (SOFC) electrolyte materials for operation at 500 °C. Solid State Ion 344:115138. https://doi.org/10.1016/j.ssi.2019.115138
20. Lee D, Lee I, Jeon Y, Song R (2005) Characterization of scandia stabilized zirconia prepared by glycine nitrate process and its performance as the electrolyte for IT-SOFC. Solid State Ion 176:1021–1025. https://doi.org/10.1016/j.ssi.2005.01.004

Algae-Assisted Microbial Fuel Cell for Electricity Generation from Food Waste

Praphulla Rao, Surabhi Vasuki, M. Nandhini, and J. Sahana

1 Introduction

Consumption of fossil fuels causes an increase in pollution which is also a major cause of global warming. Fossil fuel consumption is predicted to increase by 36% by 2030, which could make this form of energy unviable and unprofitable in the future [1]. There is a growing demand for cleaner and more renewable energy sources due to the world's population growth [2]. Researchers are seeking alternative sources of electricity for industrial and home uses since electric motors and other fuel-free alternatives have grown in popularity [3]. Food waste management is one of the major concerns in India. According to the food waste index report 2021 by the United Nation Environment Program (UNEP), in India, 90 kg/capita/year of food waste was recorded in the high-income group which was 68 and 63 in middle- and low-income groups [4]. Large quantity of food waste is disposed of in landfills without processing. In major cities, certain biological and chemical processing of food waste is carried out to reduce the load on the environment. Food waste consists of a large quantity of water. Food waste is rich in carbohydrates, lipids, and proteins that can be essentially utilized to generate energy. Anaerobic digestion is one of the old and matured techniques that is used to process food waste and generate biogas. Some of the disadvantages of anaerobic digestion are high operating cost, long solid retention time, high carbon footprint from biogas, pre-treatment of food waste, and inhibitory action of ammonia-based salts [5].

Microbial fuel cells (MFCs), which directly convert organic matter to electric energy without the need for intermediate stages, have recently been the subject of substantial research. The biodegradation efficiency of organic matter and the electron transfer efficiency have a significant impact on the generation of bioelectric energy in MFCs. In fact, appropriate anodic respiring bacteria on the anode can further

P. Rao (✉) · S. Vasuki · M. Nandhini · J. Sahana
Department of Biotechnology, B.M.S College of Engineering, Bangalore, Karnataka, India
e-mail: praphullarao.bt@bmsce.ac.in

S. T. Aruna et al. (eds.), *Recent Advances in Electrochemical Science and Technology*, Springer Proceedings in Materials 47, https://doi.org/10.1007/978-981-97-3958-5_2

improve or enhance the performance of MFCs [6]. Xin et al. [6] demonstrated the working of hydrolysis and MFC setup through which they obtained the energy of 0.245 kWh per kg of food waste. A typical dual-chambered MFC comprises an anode and a cathode chamber where oxidation and reduction occur, respectively. Microbial fuel cells provide a promising technology to handle the continuous exhaustion of fossil fuels and the steady increment in the cost of fuels. Microbial fuel cells can be broadly classified as mediator microbial fuel cells and mediator-free microbial fuel cells based on the ability mode of electron transfer. Mediators are used for facilitating the transfer of electrons to the electrode for some bacterial species that are electrochemically inactive. Chemical mediators like thionine, methyl viologen, methyl blue, humic acid, and riboflavin transfer the electrons. Studies have shown that MFCs can be used to generate electricity from organic wastes such as wastewater, sediments, and even rhizo deposits [7]. Microbial fuel cells have various applications, not only as a renewable energy source but also in bioremediation and implantable bioelectronic devices [8].

1.1 Anode Chamber

Anodic Chamber contains anaerobic electrogenic bacterial species, these species transfer electrons outside of their cells by either a direct electron transfer or a mediated electron transfer. In direct means of transfer, they use conductive cytochrome proteins, and some bacterial species form conductive pili-like appendages that further help in the formation of a biofilm on the electrode surface. Exocellular electron transfer (EET) describes the ability of certain microorganisms to generate and transfer electrons outside of their own cell. Some of the mechanisms used by bacteria for exocellular electron transfer include cellular respiration using solid metal oxides, such as iron, and possible cell–cell communication. In cellular respiration, many strains of bacteria can release electrons from a terminal oxidase in the respiratory chain to Fe (III) outside the cell, producing soluble Fe (II). Additionally, cells can also transfer electrons directly to another cell without the need for intermediates. *Pelotomaculum thermopropionicum* was observed to be linked to another type of bacteria called *Methanothermobacter thermautotrophicus*, this link was made through an electrically conductive appendage, which means that the two bacteria were able to transfer electrons directly to each other. This kind of transfer of electrons between different types of bacteria is known as interspecies electron transfer. This mechanism helps us understand how different types of bacteria can work together to generate electrical currents in microbial fuel cells. By studying the mechanisms behind interspecies electron transfer, a more efficient and effective way of generating electricity using bacteria can be developed [9]. *Geobacter sulfurreducens* are metal-reducing proteobacteria that have nanowires, they require acetate as substrate. The outer surface c-type cytochromes of *G. sulfurreducens* plays an important role in biofilm formation on electrodes [10]. In many electroactive microorganisms such as *S. oneidensis* MR-1, *G. sulfurreducens*, *T. ferriacetica*, *T. potens* strain JR, *R.*

palustris TIE-1, and *S. lithotrophicus* ES-1, the electrons produced from cellular metabolisms are transferred across the cell surface from a conductive pathway by the multiheme cytochromes. These proteins have a wide range of reduction potentials due to the presence of heme [11].

Acetate and glucose are the most frequently used substrates in microbial fuel cells. The acidity of the anolyte is an important factor that influences the growth of bacteria in the anodic chamber, acetate is rich in carbon resources and is the end product of many metabolic pathways and when acetic acid dissolves in water, it dissociates into acetate ions and hydrogen ions [12]. The most optimum pH that the microbial fuel cells are operated remains mostly at a neutral range, as a significant change in pH can shift the redox potential of the electrode, affecting the thermodynamics of the electron transfer reactions and leading to altered current generation. Sugar substrates relatively produce less acidic by-products thereby resulting in a slower decrease of anolyte pH. Bacteria such as *Shewanella putrefaciens, Aeromonas hydrophila*, and *Geobacter sulferreducens* act as efficient electrogenic microbes for mediator-free systems.

Food waste can be used as the best resource for MFCs due to its organic and nutrient-rich composition [13]. Food waste is considered a sustainable resource for microbial fuel cells because it is rich in organic compounds like carbohydrates, lipids, and proteins which act as a source for the microorganisms in order to produce protons and electrons. A large quantity of food waste is continuously generated every day and it is readily available hence it can be convenient as a sustainable feedstock for the MFCs and simultaneously helps to minimize food waste by converting chemical energy to electrical energy [14]. Organic substances present in the anode serve as the source for the bacteria. The bacteria digest this by oxidation reaction and produce electrons and using the food waste having more content of sugar can increase the electric output of MFC [15]. The type of organic substances used in the anode is also one of the criteria to increase the electric output. Food waste containing oranges yields a higher voltage output which is followed by bananas and mango. The presence of higher concentrations of sugar and organic acids helps in increasing microbial activity which in turn helps to increase the voltage output [13] (Table 1).

1.2 Cathode Chamber

In the cathode chamber, protons and electrons are combined with oxygen, to produce water. The cathode chamber must be filled with some dissolved oxygen accepter to facilitate the oxygen reduction reaction. Further, oxygen or air needs to pump at a slow circulation rate for the cathodic reaction to occur. Some of the compounds such as ferricyanide, and manganese oxide are added to the cathode chamber to enhance the kinetics of the oxygen reduction reaction. Algae are organisms that are capable of photosynthesis, depending on the cell type they can be classified as microalgae and macroalgae. By utilizing algae at the cathode chamber as substrate as opposed to a typical bacterial MFC, the dissolved oxygen in the cathode acts as an electron

Table 1 Comparison of different types of fruit waste with their maximum voltage

Sl. no	Type of fruit waste	Max. voltage (mV)	Max. power density (mA/m^2)	Reference
1	Orange	357	195	Rahman et al. [13]
2	Rambutan	170	39.94	Yaqoob [16]
3	Banana	286	78.2	Rojas-Flores et al. [17]
4	Langsat	100	23.49	Yaqoob [16]
5	Mango	180	42.29	Yaqoob [16]
6	Papaya	140	30.45	Rojas-Flores et al. [17]
7	Blueberry	240	63.11	Rojas-Flores et al. [17]

acceptor and can also be used for generating algal biomass simultaneously using it for carbon sequestration. The use of algae in the cathode chamber of an MFC eliminates the requirement for an external power source since algae produce oxygen by photosynthesis reaction so there is no need for any external source to produce oxygen. Through photosynthesis, the algae in the cathode chamber convert water and carbon dioxide to produce oxygen in the presence of sunlight. Oxygen then diffuses into the MFC system and serves as the electron acceptor at the cathode in an MFC. Hence algae-assisted MFC reduces the usage of chemical oxidizers or an external power source. The algal biomass produced in the cathode chamber helps in enhancing the voltage output of MFC. Algae help in the production of biodiesel through the transesterification process and it can also be used as a nutritional supplement in animal feed since many of the algae species are rich in proteins, amino acids, and other bioactive compounds. Algae biomass can also be used as organic fertilizer for farming since it is rich in nutrients including potassium, phosphorus, and nitrogen. The algae also facilitate carbon dioxide fixation through photosynthesis. The cathode chamber of the algae-assisted MFC acts as a bioreactor to produce algal biomass by maintaining optimized conditions of pH, temperature, light, carbon dioxide, and nutrient supply [18].

1.3 Electrode and Separator

Electrode material utilized is one of the major factors contributing to the performance of an MFC. One of the main problems in MFC is their low power output, and the major influence on output is the electrode material and size. The cost, biocompatibility, resistance, electronegativity, stability, and efficacy of electrode materials affect their capacity to give a higher output. Carbon-based materials are most significantly used for microbial fuel cell applications. Despite being good conductors some metal-based electrodes have comparatively lesser biocompatibility and a higher possibility of corrosion. Graphite, graphene, carbon felt, carbon cloth, and carbon brush are commonly used in MFC. These carbon materials are thermally activated or coated with nanometal and metal oxides to improve the power density of the MFC. Porous or mesh-like electrode materials can be used in anode for facilitating biofilm formation. However, higher biofilm formation can also result in higher resistance for more electron transfer and hence a lower output. Natural biomaterials can also be utilized for electrodes, by carbonization of various biomass.

The anode and cathode can be separated by a proton exchange membrane (PEM) or salt bridge. Nafion membrane from Dupont is commonly used in PEM, which has good thermal stability, ion selectivity, and conductivity, but it is very expensive. Apart from that, Nafion membranes are delicate and undergo biofouling by ammonium ions. Nafion has good proton conductivity, enabling efficient proton transport while reducing fuel and oxidant species crossing between the anode and cathode compartments. Nafion maintains its chemical stability in abrasive conditions, such as hot temperatures and acidic or alkaline environments. It withstands deterioration and keeps performing for a long time [19, 20]. Protons are the only ions and gas molecules that are selectively permeable across Nafion membranes, which also block the passage of most other ions. This keeps the protons' concentration gradient across the membrane, which is necessary for the creation of electrical potential in the cell.

The performance of the Nafion membrane in MFCs can be greatly impacted by biofouling caused by ammonium ions, more specifically by the buildup of ammonium ions on the membrane surface. It is well known that ammonium ions bind to Nafion, decreasing its proton conductivity and obstructing the movement of ions across the membrane [21]. As a result, the MFC's power output is decreased and its internal resistance is raised. In order to reduce ammonium ions buildup, it is crucial to manage biofouling issues, such as by implementing pre-treatment processes or optimizing the MFC working parameters. To improve the performance and reduce problems with biofouling and water management, the Nafion membrane is pre-treated before being used in MFCs. Before assembly into the MFC, the membrane is immersed in a suitable aqueous solution to ensure adequate hydration. This process aids in maintaining the necessary amount of water within the membrane and promotes proton conduction. Nafion membranes are frequently cleaned to get rid of pollutants and impurities that can compromise their performance. The two cleaning techniques that are employed are rinsing with water or using particular cleaning solutions.

1.4 Algae-Assisted Microbial Fuel Cell

In this work, the dual-chambered algae-assisted MFC was built using different electrode materials, 0.5 kg of composted food waste as anode, and algae lake water as

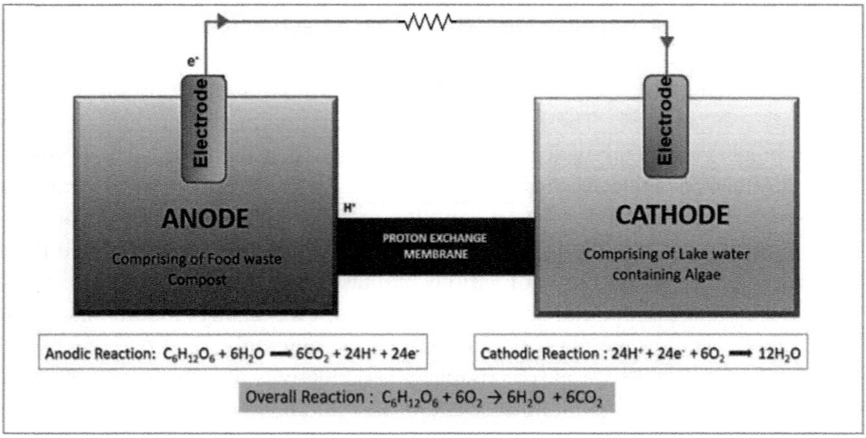

Fig. 1 Schematic of the constructed algae-assisted microbial fuel cell

cathode substrates. A schematic representation of algae-assisted MFC is given in Fig. 1.

2 Experimental

Food waste was prepared using fruit peels of orange, musambi, papaya, mango, muskmelon, and vegetable leftovers such as cabbage, tomato, radish peels, and dried flowers from the institute canteen. This food waste was made into a paste using a laboratory mixer and then mixed with soil which contains microbes. The food waste and soil were layered in alternative stacks and allowed to degrade for 3 weeks. Then the degraded semi-liquid compost was taken into the anodic chamber of MFC. The cathodic chamber was filled with algae water collected from the nearby lake. The salt bridge was constructed with PVC pipe filled with 10% agar and 10% NaCl in a 13 cm PVC pipe with a 1 cm diameter. Small scale MFC with 0.5 kg food waste in the anode and 500 mL of Algae water was constructed. A photograph of the setup is shown in Fig. 2a. Aluminum mesh was used for the electrode, having dimensions of 8.5 cm in length and 2 cm in breadth. Sparging of nitrogen gas was done intermittently to maintain anaerobic conditions in the anodic chamber. The intermittent sparging of nitrogen gas in the anode helps in removing the oxygen present in the chamber as a result of this the oxidative stress on the anaerobic microorganisms can be prevented which promotes the optimal activity and growth of bacteria [22].

Two plastic containers were used to construct the MFC. Two containers were drilled with holes at the top and sides to fit the copper wires and salt bridge respectively. The holes were sealed with a hot glue gun, and an air inlet was made for the cathode chamber. Copper wire was soldered to the aluminum mesh electrode. A multimeter was used to measure the output voltage at different time intervals of

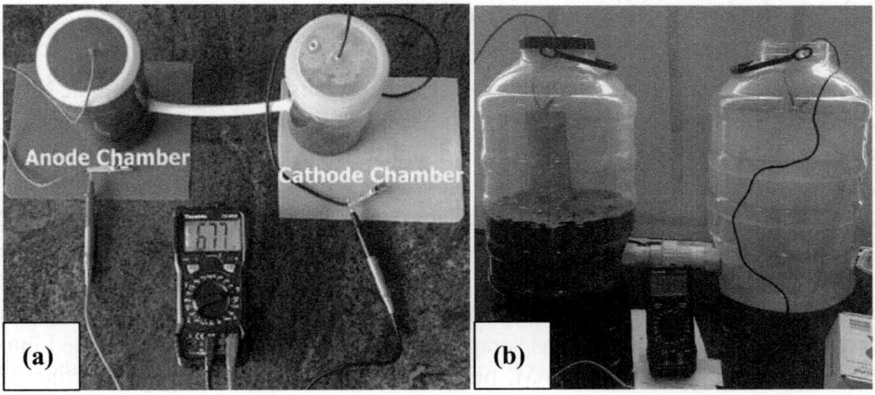

Fig. 2 MFC setups used in the laboratory: **a** MFC with 0.5 kg of food waste, **b** MFC with 5 kg of food waste

one hour, the copper wires were connected to the multimeter using alligator clips, and the readings were taken. The current was measured in series using an LED as a load. A red LED was used as a load to connect into series with the MFC apparatus to measure current and in parallel to measure voltage.

2.1 Effect of Electrode Materials

In another arrangement, a large-scale MFC was fabricated to process the food waste. The picture of the setup is shown in Fig. 2b. Two large transparent plastic containers were used for each of the half cells. Combinations of aluminum/aluminum, copper/copper, aluminum/copper, graphite/copper, and copper/graphite were compared. The combination of copper as the anode and aluminum mesh as the cathode resulted in the highest potential gradient compared to all other combinations. Aluminum mesh (75 mm × 200 mm) and resistance of 0.7 Ω were mainly chosen for its larger surface area to facilitate biofilm formation. The use of aluminum mesh in the cathode helps in the growth of algal biofilm on the surface of the electrode and also offers a lower corrosion rate compared to copper. Aluminum is also non-toxic for the growth of algae, contradictorily prolonged exposure to copper inhibits algal growth due to the toxic nature of the metal. For the anode substrate, food waste such as vegetable and fruit wastes are collected and left to decompose in soil for one month, 5 kg of food waste sludge was prepared in a similar manner as the small-scale MFC. In the cathode, lake water containing algae was used. Here scaling up MFC was done to process the food waste from 0.5 kg to a 5 kg quantity. A Nafion membrane with a 30 mm diameter was used as a separator instead of a salt bridge to minimize the internal resistance accounted from the salt bridge. Prior to the application, Nafion membrane is soaked in deionized water and followed by sulphuric acid for 24 h.

2.2 *Isolation of Microorganisms*

The bacteria sample grown in the anodic chamber was isolated after letting the MFC function for 7 days. Since bacteria form a layer of biofilm on the electrode surface, some portion of the sludge was collected and serially diluted. The serially diluted sample was grown on nutrient agar media containing peptone, sodium chloride, yeast extract, and agar. Since these exoelectrogenic bacterial species are anaerobic, the isolation should be done by maintaining an oxygen-free environment. To isolate the microorganisms growing over the anodes, the surface of the electrode was rinsed with a stream of sterile distilled water until they were free of visible debris. The first millimeter of the anode was scraped with a sterile razor blade and added into 1.5 mL phosphate buffer, 50 mM, pH 7.2, producing a suspension consisting of electrode-associated microbes. The obtained suspension was serially diluted up to 10–6 and the plates were incubated in both anaerobic and aerobic conditions. A candle anaerobic jar was used to maintain anaerobic conditions and the petri dishes were incubated for 24 h. The algae were isolated from the lake water samples, grown using BG-11 media and incubated in light and dark conditions for 12 h each. Algae were morphologically characterized and observed under a microscope. Algal cells from different regions of the cathode chamber were taken for further growth and were added back into the cathode chambers. The pH of the catholyte was measured using a digital pH meter intermittently every day (Thermo Scientific Orion star A100 series). The dissolved oxygen level was measured using a titrimetric estimation following the Winkler method [23] and BOD was measured after incubating for 5 days. The reducing sugar content in the sludge of the anode chamber was estimated using a qualitative Benedict's test and was confirmed by performing the quantitative colorimetric estimation of glucose using the DNS method, using glucose as the stock solution to obtain a standard graph.

2.3 *Connection of MFCs in Series*

Three MFC setups were constructed following the same procedure as discussed previously and were connected in series; two transparent containers connected via a salt bridge of 10% concentration were prepared. Aluminum and copper were used at the cathode and anode respectively based on the results obtained from the comparison of different electrode materials in large-scale MFC. The circuit diagram representing the series connection of three MFC is shown in Fig. 3. Each MFC contains 0.5 kg of food waste compost and 500 mL of algae water.

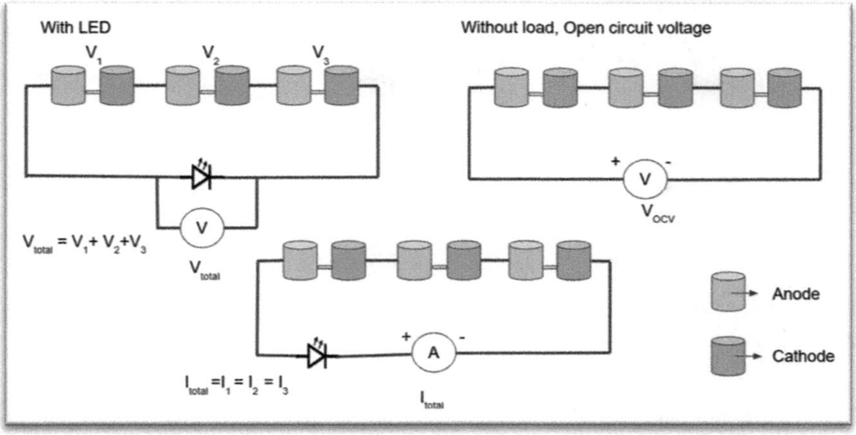

Fig. 3 Circuit diagram representing the series connection of three MFC, each with 500 mL capacity

3 Results and Discussions

The small-scale MFC having 0.5 g of food waste was operated for a period of 6 months without the addition of any extra nutrients. The voltage and the current characteristics were monitored frequently for a period of 30 days where the continuous readings were observed with the help of a multimeter. The maximum voltage output was 786 mV and the current of 200 μA was observed on the 11th and 12th day. Figure 4 shows the voltage characteristics of MFC with 0.5 kg of food waste and 500 mL of algae water. The trend of the graph represents during the initial days of MFC operation, the voltage was increasing gradually till day 12, and then it slowly started to decrease this is because of a decrease in the organic matter. Since the microbes utilize the organic matter for electron and proton production, as the organic matter decreases the bacteria don't get enough amount of food which results in the drop of voltage and current. The voltage characteristics were continuously monitored in 1-h intervals (Fig. 4b) at the beginning of the MFC operation and there was a voltage drop during the mornings when compared to the readings obtained during the afternoon. This may be due to the production of less oxygen during the night due to the absence of photosynthesis. Since the light and dark cycles influence the oxygen production, the growth rate depends on it and also, the anaerobic bacteria in the anodic chamber take time to digest the substrate and grow. It can be observed from the trend of the graph that the value of voltage is obtained higher during mid-day due to more photoperiod exposure to the algae chamber. An increase in algae growth was observed due to the formation of biofilm and coloration in the cathode chamber.

The MFC was scaled up from 0.5 to 5 kg where the salt bridge was replaced by the Nafion membrane. Scaling up of the MFC setup has resulted in an increased voltage output due to the presence of more anaerobic bacteria leading to more electron transfer and the use of the Nafion membrane.

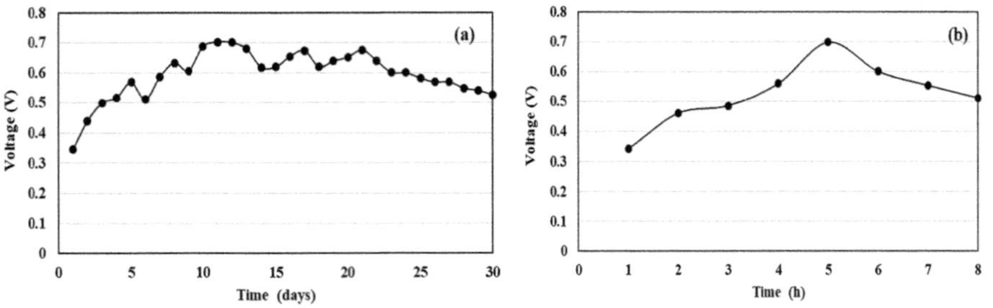

Fig. 4 Voltage characteristics of MFC with 0.5 kg of food waste and 500 mL of algae water, **a** for 30 days, **b** for 8 h

The voltage characteristics for the different electrode materials were measured for the large-scale MFC in which 5 kg of food waste was taken. Figure 5a gives the 8th day voltage of MFC with different combinations of electrodes. The maximum voltage value of 1.061 V was observed on the 8th day for the aluminum and copper electrode combination, where the aluminum electrode was used in the anode chamber and the copper electrode in the cathode. The aluminum and the platinum electrode combination also showed the highest result of 0.923 V on the 4th day. Overall the best results were obtained for the Aluminum and Copper combination, so the same combination was maintained in the large-scale MFC. The open circuit voltage profile of large-scale MFC for 20 days is given in Fig. 5b.

The voltage output observed from scaling up was not as substantial, so three MFCs were connected in a series of results were compared. During the initial day's low voltage output was recorded and then there was an increase in the trend and again it dropped down slowly. The maximum voltage obtained from the series connection was 2.16 V on the 19th day (Fig. 6a). The voltage was measured for a period of 30 days and it was giving a voltage in the range of 1.5 V. The current from the series setup was found to be 10 mA which was measured using a 1 Ω standard resistor (variable resistor) and the galvanometer.

Single MFCs were allowed to function continuously for 4 months and open circuit voltage was monitored (Fig. 6a). MFC with 0.5 kg food waste and salt bridge showed

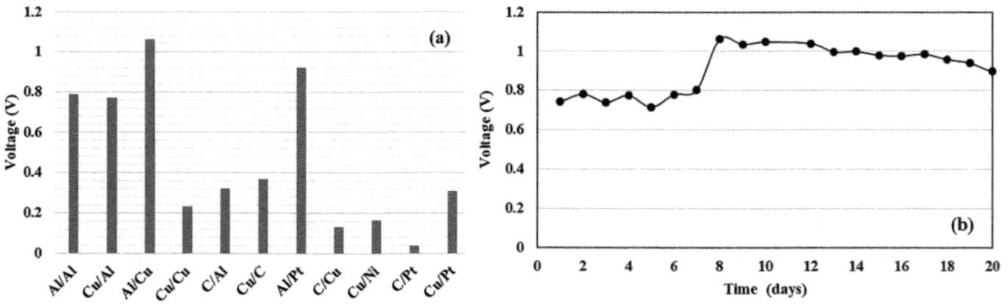

Fig. 5 Voltage characteristics of MFC with 5 kg of food waste, **a** different electrode materials **b** day-wise voltage monitoring of MFC with Cu as anode and Al as cathode

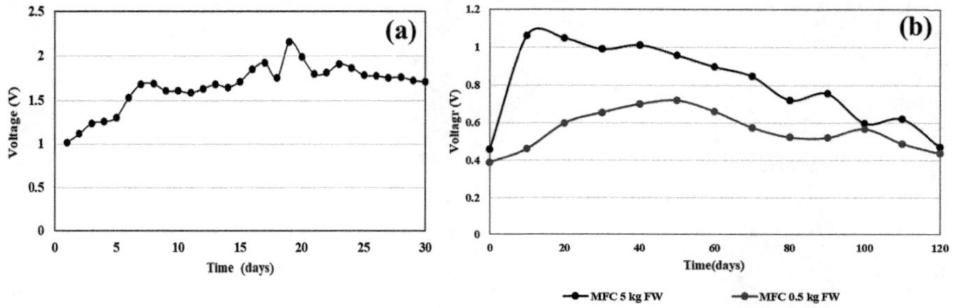

Fig. 6 Voltage characteristics **a** MFCs connected in series, **b** working of single MFC for 4 months

37% depletion in voltage. MFC with 5 kg food waste and Nafion membrane showed 58% depletion in voltage. This can be due to the higher internal resistance in MFC with 5 kg food waste. The maximum voltage obtained for the 0.5 kg of food waste was 0.700 V and the maximum voltage obtained for the 5 kg of food waste was 1.06 V. Scaling up of MFC to 5 kg resulted in increasing the voltage compared to the 0.5 kg MFC since the quantity of food waste was large, so more number of bacteria helped in electron transfer and also usage of Nafion membrane facilitated the efficient transfer of protons. The open circuit voltage increases when multiple MFCs were connected in a series manner here the individual cell voltages are summed up to produce a higher voltage [24]. Over a period of time the voltage value reduces since the bacteria uses the organic substance to produce the electrons after some time there will be depletion in the nutrient source and also accumulation of metabolic by-products leads to a decrease in the voltage over a long term [25]. Various factors that influence the shelf life of the MFCs are the type of electrode used, the type of food waste and the proton exchange membrane, and the proper maintenance of anaerobic conditions. It is essential to select the proper electrode materials that offer less corrosion and have less internal resistance, periodically adding fresh food waste to the anode, and maintaining anaerobic conditions [26].

The presence of algae increases the dissolved oxygen in the system through photosynthesis. As the biological oxygen demand and dissolved oxygen (DO) are inversely proportional to each other, an increase in the DO results in a decrease in the BOD. The DO and the BOD were measured by the Winkler method where the DO of the cathode which has a lake water sample containing algae was found to be 5.9 mg/L and the BOD is 12.67 mg/L. So here the BOD is higher than the DO which indicates that less oxygen is being removed from the water and it has higher organic matter. The dissolved oxygen in the water is consumed by microorganisms to decompose the organic matter in the water.

Morphological characteristics of microorganisms were done using a laboratory microscope. Figure 7a shows aerobic bacteria which was grown in 10^{-4} dilutions. It has many different colonies since the anodic sludge consists of different co-cultures. Figure 7b is the anaerobic culture which was grown in a petri dish inside the anaerobic jar. There are not many colonies as compared to the aerobic culture.

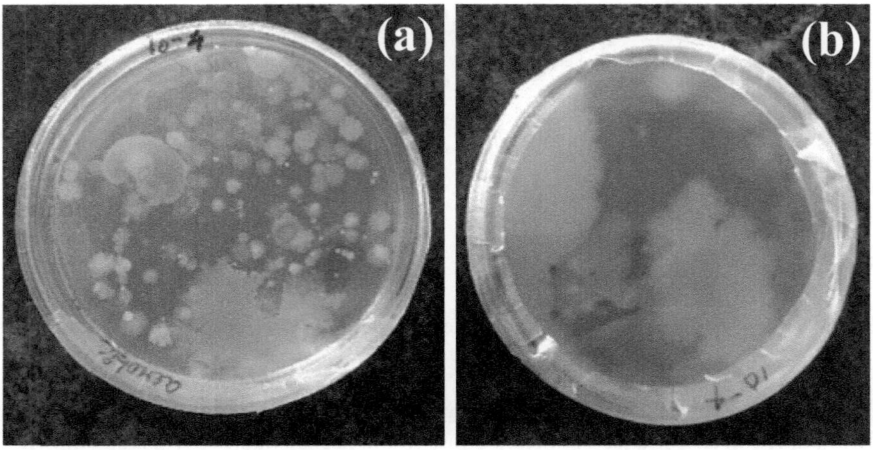

Fig. 7 Growth of isolated bacteria in petri dish **a** aerobic bacteria, **b** anaerobic bacteria

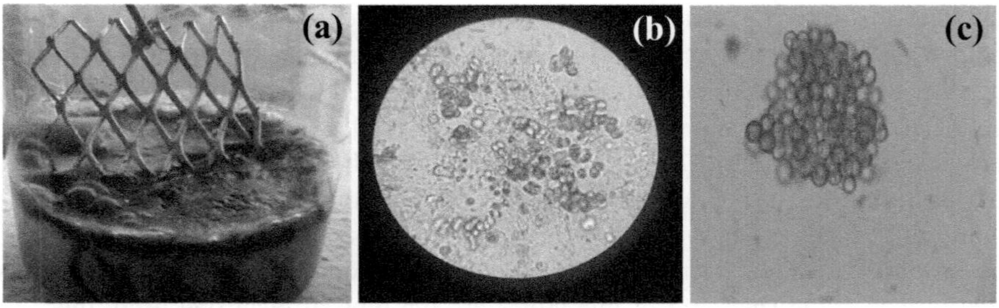

Fig. 8 Images of algae from the cathodic chamber

Algae was sourced from Yediyur Lake and grown in the cathodic chamber of MFC, it contained many blue-green algal species, according to Asulabha et al. [27] Yediyur Lake was found to contain mostly *cyanophyceae* species out of which the most predominant species was *Microcystis* sp. This is due to a neutral or slightly alkaline pH of 7.4–8.5. As characterized under the microscope (Fig. 8b, c) and from comparison and growth of algae in BG-11 media we could say that a few species contained in the sample would probably be of Chlorella sp. and Microcystis sp.

4 Conclusions

MFC serves as an alternating energy source to solve the energy crisis and environmental pollution. MFC development makes it possible to generate electricity as well as to remove pollutants from the water and it also helps in reducing food waste. The performance of the MFC mainly depends on the kinetics of the electrode reactions.

Many different combinations of electrode materials have been tested to enhance the performance of MFC. The aluminum and copper combination showed a good result which can be used for anode and cathode construction. Because of their good chemical stability, accessibility, and electronegativity, they can be the best alternative even in terms of cost, also they are cheap as we are using very thin sheets of copper and aluminum which helps to increase biomass growth. The pH of the algae was increasing as the biomass growth increased which facilitates the acceptance of more electrons and increases the overall output in the MFC. The maximum voltage and current output of 786 mV and 200 μA were obtained respectively on the 11th and 12th day for the MFC with 0.5 kg food waste. The MFC with 5 kg waste yielded a maximum voltage of 1.061 V on the 10th day with a current of 20 μA. The series-connected MFC increased the cell voltage compared to individually operated cells. The different voltage values were recorded for three of the individual setups that were connected in a series manner, the three 500 mL setups which had the same quantity of food waste sludge was showing different voltages as 1.212 V, 0.302 V, and 0.850 V, respectively. This may be due to varied microbial activity in the individual cells. Voltage was above 1.5 V throughout 30 days in the series connection and a maximum of 2.16 V was obtained on the 19th day of operation. MFC could work efficiently over two months with a 15% decrease in voltage. After 8 months of working degradation of potential was around 80%. The voltage variation depends on the type of food waste used, the salt bridge composition, and also the size of the electrode. The shelf life of the MFC can be 5–10 months it can be improved by adding redox mediators, once their shelf life is over, the anodic sludge can be used as manure. The series combination of MFC showed better results compared to the other setups.

References

1. Martins T, Barreto AC, Souza FM, Souza AM (2021) Fossil fuels consumption and carbon dioxide emissions in G7 countries: empirical evidence from ARDL bounds testing approach. Environ Poll 291:118093. https://doi.org/10.1016/j.envpol.2021.118093
2. Gielen D, Boshell F, Saygin D, Bazilian MD, Wagner N, Gorini R (2019) The role of renewable energy in the global energy transformation. Energy Strat Rev. 24:38–50. https://doi.org/10.1016/j.esr.2019.01.006
3. Palanisamy G, Jung H, Sadhasivam T, Kurkuri MD, Kim SD, Roh SH (2019) A comprehensive review on microbial fuel cell technologies: Processes, utilization, and advanced developments in electrodes and membranes. J Cleaner Prod 221:598–621 (2019). https://doi.org/10.1016/j.jclepro.2019.02.172
4. Sinha S, Tripathi P (2021) Trends and challenges in valorization of food waste in developing economies: a case study of India. Case Stud Chem Environ Eng 4:100162. https://doi.org/10.1016/j.cscee.2021.100162
5. Wenjing L, Chenxi J, Junting Z, Junqing X, Dianhai Y, Guangming L (2021) Conversion of food waste into biofuel and biocarbon. In: Khan A, Jawaid M, Pizzi A, Azum N, Asiri AM, Isa IMD (eds) Advanced technology for the conversion of waste into fuels and chemicals, vol 1, pp 383–449. ISBN: 9780128235270

6. Schamphelaire LD, Bossche LV, Dang HS, Höfte M, Boon N, Rabaey K, Verstraete W (2008) Microbial fuel cells generating electricity from rhizodeposits of rice plants. Environ Sci Technol 42:3053–3058. https://doi.org/10.1021/es071938w
7. Katz E (2015) Implantable biofuel cells operating in vivo—potential power sources for bioelectronic devices. Bioelectronic Med 2:1–12. https://doi.org/10.15424/bioelectronmed.2014.00011
8. Wang F, Gu Y, O'Brien JP, Yi SM, Yalcin SE, Srikanth V, Shen C, Vu D, Ing NL, Hochbaum AI, Egelman EH, Malvankar NS (2019) Structure of microbial nanowires reveals stacked hemes that transport electrons over micrometers. Cell 177:361–369. https://doi.org/10.1016/j.cell.2019.03.029
9. Logan BE (2009) Exoelectrogenic bacteria that power microbial fuel cells. Nat Rev Microbiol 7(5):375–381. https://doi.org/10.1038/nrmicro2113
10. Kiran EU, Trzcinski AP, Ng WJ, Li Y (2014) Bioconversion of food waste to energy: a review. Fuel 134:389–399. https://doi.org/10.1016/j.fuel.2014.05.074
11. Paquete CM, Morgado L, Salgueiro CA, Louro RO (2022) Molecular mechanisms of microbial extracellular electron transfer: the importance of multiheme cytochromes. Front Biosci 27(6):174. https://doi.org/10.31083/j.fbl2706174
12. Kt O, Onyeaka H, Meyer E, Nwokolo N (2021) Microbial fuel cells, a renewable energy technology for bio-electricity generation: a mini-review. Electrochem Commun 125:107003. https://doi.org/10.1016/j.elecom.2021.107003
13. Rahman W, Suzana Y, Mohammad S (2021) Screening of fruit waste as substrate for microbial fuel cell (MFC). AIP Conf Proc 2332:020003. https://doi.org/10.1063/5.0043556
14. Zafar H, Peleato NM, Roberts DJ (2022) A review of the role of pre-treatment on the treatment of food waste using microbial fuel cells. Environ Technol Rev 11:72–90. https://doi.org/10.1080/21622515.2022.2058426
15. Khoirunnisa N, Anwar S, Sudadi U, Santosa DA (2021) Effect of organic substrate type in electricity production from microbial fuel cell (MFC) inoculated by Staphylococcus saprophyticus, ICBB 9554. IOP Conf Ser Earth Environ Sci 927012028. https://doi.org/10.1088/1755-1315/927/1/012028
16. Yaqoob A (2022) Local fruit wastes driven benthic microbial fuel cell: a sustainable approach to toxic metal removal and bioelectricity generation. Environ Sci Poll Res 29:32913–32928. https://doi.org/10.1007/s11356-021-17444-z
17. Rojas-Flores S, Benites SM, Cruz-Noriega M, Cabanillas-Chirinos L, Valdiviezo-Dominguez F, Álvarez MA, Vega-Ybañez V, Angelats-Silva L (2021) Bioelectricity production from blueberry waste. Processes 9:1301. https://doi.org/10.3390/pr908130
18. Nagendranatha Reddy C, Nguyen HT, Noori MT, Min B (2019) Potential applications of algae in the cathode of microbial fuel cells for enhanced electricity generation with simultaneous nutrient removal and algae biorefinery: Current status and future perspectives. Bioresource Technol 122010. https://doi.org/10.1016/j.biortech.2019.122010
19. Yaqoob AA, Ibrahim MNM, Umar K (2021) Electrode material as anode for improving the electrochemical performance of microbial fuel cells. Energy Storage Battery Syst. https://doi.org/10.5772/intechopen.98595
20. Kannan N, Donnellan P (2021) Algae-assisted microbial fuel cells: a practical overview. Biores Technol Rep 15:100747. https://doi.org/10.1016/j.biteb.2021.100747
21. Zabed H, Amru Nasrulhaq B, Jaya S, Faruq G (2016) Evaluation of the quality of dried distiller's grains with solubles for normal and high sugary corn genotypes during dry–grind ethanol production. J Cleaner Prod 142:4282–4293. https://doi.org/10.1016/j.jclepro.2016.11.180
22. Bavasso I, Montanaro D, Petrucci E, Palma LD (2018) Shortcut biological nitrogen removal (SBNR) in an MFC anode chamber under microaerobic conditions: the effect of C/N ratio and kinetic study. Sustainability 10:1062. https://doi.org/10.3390/su10041062
23. Carvalho A, Costa R, Neves S, Oliveira CM (2021) Determination of dissolved oxygen in water by the Winkler method: performance modelling and optimisation for environmental analysis. Microchem J 165:06129. https://doi.org/10.1016/j.microc.2021.106129

24. Kim B, An J, Kim D, Kim T, Jang JK, Lee B, Chang IS (2013) Voltage increase of microbial fuel cells with multiple membrane electrode assemblies by in series connection. Electrochem Comm 28:131–134. https://doi.org/10.1016/j.elecom.2012.12.019
25. Oh S, Logan BE (2007) Voltage reversal during microbial fuel cell stack operation. J Power Sources 167:11–17. https://doi.org/10.1016/j.jpowsour.2007.02.016
26. Huang S, Zhang J, Pi J, Gong L, Zhu G (2021) Long-term electricity generation and denitrification performance of MFCs with different exchange membranes and electrode materials. Bioelectrochemistry 140:107748. https://doi.org/10.1016/j.bioelechem.2021.107748
27. Asulabha KS, Sincy V, Jaishanker R, Ramachandra TV (2018) Algal diversity in urban lakes of vrishabhavathi valley, greater Bangalore. In: Proceedings: the 11th biennial lake conference, conference on conservation and sustainable management of riverine ecosystems

Design and Performance Studies of Supercapacitor Coin Cells from Hierarchical Activated Carbon Acquired from Stems of *Musa Acuminata*

V. Tejashree, M. G. Radhika, S. Srividhya, and Kathyayini Nagaraju

1 Introduction

It is well known that the increase in population has created multifaceted demands globally and meeting these demands has resulted in a massive imbalance in the ecosystem [1]. As a consequence of this, there is a tremendous upsurge in the production of various types of crops. Considering the case of bananas, it is one of the most widely cultivated, consumed, and traded fruits worldwide. An estimate of about 19.6 million tons of bananas were exported globally in the year 2022 [2]. Due to the above reason large quantities of plant waste or biowaste are generated and hence waste management has become a major concern that requires immediate resolution. World-wide issues like waste management and energy crisis should be catered to in a cohesive manner that brings us a step closer toward sustainable solutions. Additionally, the depleting fossil fuels have also inevitably challenged the scientific community to foster new technologies for developing alternate energy storage devices [3]. Currently, four different categories of energy storage devices are available for diverse small and large-scale applications, namely, electrochemical, electrical, chemical, and mechanical. Electrochemical capacitors, electrochemical supercapacitors, batteries, and fuel cells are the popular electrochemical energy storage systems [4, 5].

A device that is capable of storing electrical energy in the form of an electrical double layer (EDL) that forms at the interface of an electrolyte solution and electronic conductor is recognized as an electrochemical capacitor [6]. Supercapacitors are a special type of capacitors wherein they have high capacitance value and energy density compared to traditional capacitors. The prefix "Super" is used because they

V. Tejashree · S. Srividhya · K. Nagaraju (✉)
Center for Incubation Innovation Research and Consultancy (CIIRC), Joythy Institute of Technology, Tataguni, Bengaluru 560082, India
e-mail: nkathyayini45@gmail.com

M. G. Radhika
Department of Physics, RV College of Engineering, Bengaluru 560059, India

S. T. Aruna et al. (eds.), *Recent Advances in Electrochemical Science and Technology*, Springer Proceedings in Materials 47, https://doi.org/10.1007/978-981-97-3958-5_3

have high specific power and specific energy. Supercapacitors can be employed independently or in combination with other types of energy storage devices for better outcomes. Supercapacitors are utilized in many consumer electronics, backup power supplies, portable electronic devices, electric vehicles, and hybrid vehicles [4, 7, 8]. Supercapacitors have such varied applications due to their long cycle stability [>5 lakhs], high power density, and quick charge duration compared to batteries [9].

Carbon-based compounds are incredibly versatile because of their attractive properties such as (a) adjustable porosity, (b) high conductivity, (c) controllable heteroatom doping, (d) lightweight, (e) low cost, (f) easy processability in the synthesis of composites, (g) increased specific surface area (SSA), (h) pore volume (PV) and (i) pore size distribution (PSD). Due to the amalgamation of all these properties, carbon-based compounds hold enormous potential for emerging technologies and sustainable energy innovations [8, 10]. As we converge to the biomass/biowaste-derived carbons, it is evident that the ready availability, environmentally friendly, and renewability aspects are what make the usage of biomass all the more interesting [11]. Especially activated carbons (physical or chemical activation) obtained from biomass are being used in electrochemical double layer capacitors (EDLC) supercapacitors. From the cited literature, it has been reported that Shaker et al. [11] produced activated carbon from banana stems by a combination of chemical and physical activation, where chemical activation was carried out using mild concentrations (0.3–0.7 M) of KOH and physical activation using CO_2 gas at 850 °C. This synthesized activated carbon demonstrated a specific capacitance (Cs) of 170 F/g with 0.5 M KOH electrolyte [12]. In their further work, the authors used 0.5 M $ZnCl_2$ as the chemical activating agent along with CO_2 gas at 900 °C for physical activation of banana stem fibers which exhibited a Cs of 179 F/g in cyclic voltammetry studies [13]. Dulyaseree et al. [14] carbonized banana stems at 400 °C for optimum results and activation was carried out by mixing NaOH with the carbon mass in a 1:2.5 ratio which displayed a specific capacitance of 55 F/g at 5 mV/s. However, in the present work the activated carbon derived from banana stem has exhibited better specific capacitance (Cs) values in, cyclic voltammetry (CV), and chronopotentiometry (CP) which are provided and discussed in detail further in the text.

It is well established that chemical activation is the most effective method for producing porous carbon because of the tangible benefits such as low activation temperature, time, and good yield. KOH as a chemical reagent is used to obtain high specific surface area (SSA) and pore volume (PV). KOH aids in creating defined micropores and mesopores thereby facilitating the formation of porous carbon [10]. Reduction of KOH via a series of redox and decomposition reactions generates the nano-sized pores [15]. In previous work, the weight ratio of the reactants—carbon:KOH as 1:4 has been optimized by considering all the activation parameters [16]. This study demonstrates the potential of KOH-activated carbon from stems of banana plant as a potential EDLC material in both three-electrode system and in coin cells using 1 M KOH as the electrolyte. Further, the practical application of two symmetric coin cells connected in series to power a 1-W LED has been demonstrated.

2 Experimental

2.1 Materials and Methods

The banana plant stem was procured from an agricultural field in Karnataka, India. The chemicals like potassium hydroxide (KOH), sodium hydroxide (NaOH), hydrochloric acid (HCl), polyvinylidene difluoride (PVDF), N-methyl pyrrolidone (NMP), and isopropyl alcohol (IPA) used were of analytical grade. Toray carbon paper, Ketjen black, and graphite felt were purchased from Synergy India. Separator (Whatman filter paper-41) used was from GE Healthcare Life Sciences.

2.2 Preparation of Activated Carbon

The outer layer of the banana stem was removed and the inner core was cut longitudinally and perpendicularly into smaller pieces and soaked in 0.1 N HCl. These pieces were then thoroughly washed with water to eliminate basic inorganic salts. Similarly, the material was then soaked in 0.1 N NaOH and was again washed with water to remove the acidic impurities. The material was then washed and dried at 110 °C for 24 h.

The carbonization process was carried out under optimized conditions at 450 °C for 1 h under N_2 (100 mL/min) atmosphere in a tubular furnace. The carbon mass obtained was cooled to room temperature and was ground using a mortar and pestle to obtain a fine powder. For the activation of the above-obtained carbon, 10 M KOH solution was used (Carbon:KOH::1:4) and was heated at 110 °C for 16 h in a silica crucible. This was further calcined in N_2 atmosphere (100 mL/min) at 750 °C for 1 h. The acquired lumpy carbon mass was first cooled to room temperature, then crushed to create a uniform texture, and was washed with a solution of 6 M HCl to remove excess KOH. The carbon mass was washed with water until the pH of the filtrate was neutral. At last, the carbon mass was dried at 110 °C for 16 h in a hot air oven and labeled as KCB (KOH-activated carbon from banana stems).

2.3 Characterization of Activated Carbon (KCB)

The carbon materials were characterized by different analytical techniques to study their morphological, physicochemical, and electrochemical properties. Powder X-ray diffraction (P-XRD) patterns were collected using Bruker X-ray diffractometer with Cu Kα radiations ($\lambda = 0.154$ nm) in the 2θ range of 5°–80° at 40 kV and a scan rate of 2° /min. Brunner-Emmett-Teller (BET) surface area and other surface properties were estimated from the BELSORP Max instrument. Scanning electron microscopy (SEM) images were acquired on a Gemini (Model Ultra 55) scanning electron microscope. Transmission electron microscopy (TEM) images were obtained using a Tecnai 20 transmission electron microscope.

2.4 *Fabrication of the Working Electrode and Electrochemical Studies*

For the preparation of the working electrode, 85 wt% of the carbon material with 10 wt% of ketjen black (EC 600JD) and binder (5 wt% of polyvinylidene di- fluoride in N-methyl pyrrolidone) were mixed together using an agate mortar and pestle. This slurry was then coated onto a 1 cm^2 Toray carbon paper and was vacuum dried at 80 °C for 16 h. The Toray carbon paper was coated with about an active mass of 1 mg/cm^2 of carbon material.

For examining the electrochemical properties of the carbon material in three-electrode system, a saturated calomel electrode was used as a reference electrode along with a platinum electrode as an auxiliary electrode and the synthesized carbon material was coated on Toray carbon as the working electrode. This three-electrode setup was tested in 1 M KOH solution using the CHI 660E electrochemical workstation. To study the electrochemical properties of the carbon material, cyclic voltammetry (CV), chronopotentiometry (CP), and electrochemical impedance spectroscopy (EIS) techniques were employed.

2.5 *Fabrication of Symmetric Supercapacitor Device (Symmetric Coin Cell)*

The graphite felt was soaked in ethanol and was ultra-sonicated at a frequency of 40 kHz (100-W power) in ethanol for 20 min. Then the graphite felt was dried at 100 °C for 1 h. About 3 mg of KCB was used to prepare the slurry as mentioned in the previous section and was uniformly coated on the graphite felt (3 mg KCB on each electrode) and left to dry at room temperature for 16 h. A Swagelok-type coin cell (CR-2032) was constructed by assembling the positive case, KCB coated graphite felt (the current collector), separator (Whatman filter paper-41) along with the spacer, wave spring, and negative case. A few drops of 1 M KOH solution were added and 2 tons of pressure was applied to sandwich the parts together. This coin cell was then used to conduct CV, CP, EIS, and cycle stability studies using the CHI 660E electrochemical workstation.

3 Results and Discussion

3.1 P-XRD, BET, SEM, and TEM Analysis

KOH-activated carbon obtained from banana stems was characterized by P-XRD, BET, SEM, and TEM. The X-ray diffraction pattern of KCB is provided in Fig. 1a. The XRD pattern of KCB indicated less crystallinity with diffraction peaks at $2\theta = 25°, 42°$, and $43°$ which corresponds to various (hkl) planes such as (002), (100), (101) respectively. The other diffraction peaks at $2\theta = 52°, 57°$, and $62°$ accounts for the (102), (004) and (103) planes. These 2θ values as well as the (hkl) planes correspond to the presence of hexagonal carbon according to ICDD 01-075-1621. Further, the peak at $2\theta = 35°$ could be attributed to remnant K_2O (ICDD: 00-025-0626) formed during the activation process [17].

With the aid of Brunner-Emmett-Teller (BET) surface analyzer, different surface parameters like surface area (m^2/g), pore volume (cm^3/g), and pore diameter (nm) were curated for KCB sample. The N_2 adsorption–desorption isotherm of KCB was

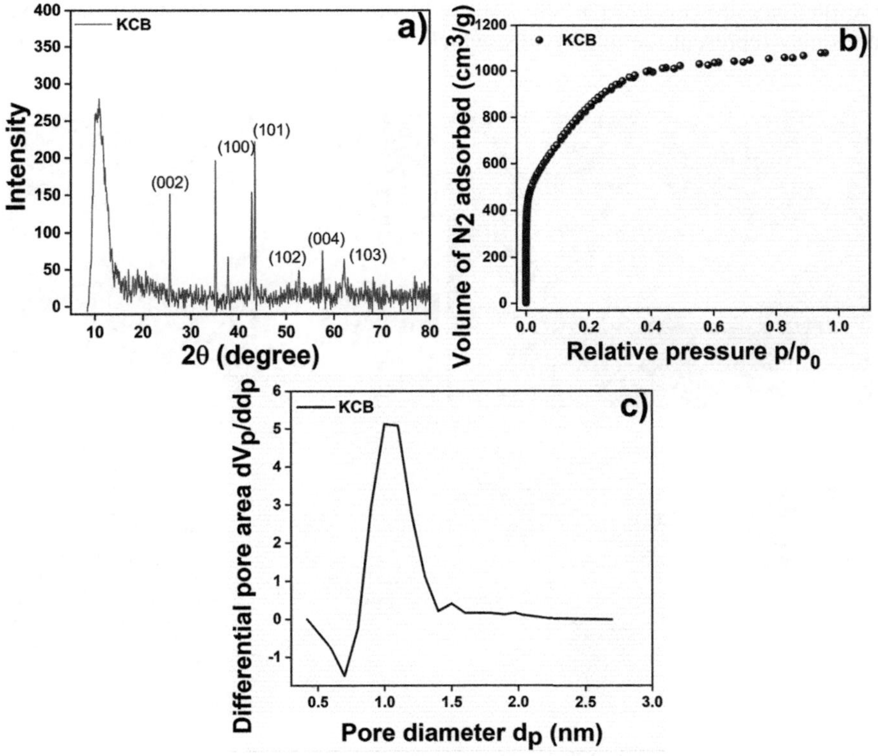

Fig. 1 **a** P-XRD graph of KCB, **b** N_2 adsorption–desorption isotherm, and **c** micropore plot (MP) of KCB

of type I (Fig. 1b), with significant N_2 adsorption at p/p_0 less than 0.01 indicating the large number of microporous structures with an average pore diameter of 1.85 nm and mesopores with an average diameter of 2.32 nm. From the BET surface area analysis of KCB, the surface area obtained was found to be 3059 m^2/g with a pore volume of 1.66 cm^3/g and an average pore diameter of 2.17 nm.

The SEM images revealed micron-scaled structures that were agglomerated or clustered in the non-activated carbon sample (Fig. 2a). The uneven terrain on the large structures showed overlapped figures with grain-like entities assembled on them. At the bottom right corner of the image, we can also witness structures that resemble tightly held clusters of particles. These observations indicate non-porosity of the carbon sample. After the activation using KOH the agglomerated structures in Fig. 2a seemed to have opened up to reveal smaller irregularly shaped structures (Fig. 2b), this may have seemed to increase the surface as compared to the image of non-activated carbon. Due to the irregularity, there are small voids/pores at the base of these structures that are attributed to the porosity of KCB.

Further, delving with TEM data of non-activated carbon in Fig. 2c, we can see globular-shaped structures that are feebly porous and in the size range of 85–150 nm. Whereas in the KOH-activated KCB (Fig. 2d) there was a complete change in morphology showing fine particles. The light minute spots in contrast with the darker ones may represent the micropores, which was substantiated by the pore diameter

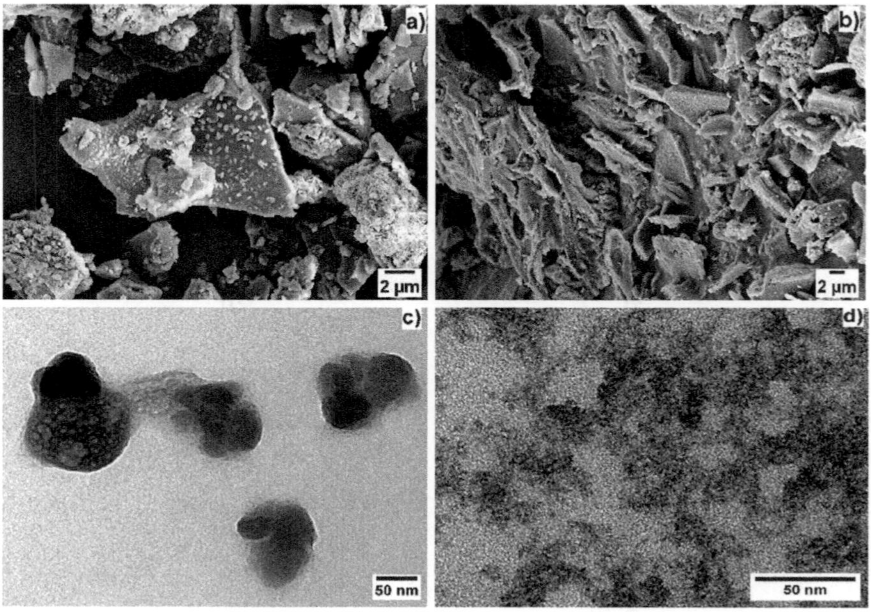

Fig. 2 **a** SEM image of non activated carbon from the banana stem, **b** SEM image of KOH-activated carbon from the banana stem (KCB), **c** TEM image of non-activated carbon from banana stem, and **d** TEM image of KOH-activated carbon from the banana stem (KCB)

value of 1.85 nm from Brunner-Emmett-Teller (BET) surface area analysis [18, 19]. The evidence of the large number of micropores has been inevitably reflected in the large surface area (3059 m^2/g) of the KCB material.

3.2 Electrochemical Performance Measurements of KCB in Three Electrode System

The electrochemical performance of KCB was analyzed in a three-electrode setup using 1 M KOH solution as the electrolyte for all cyclic voltammetry (CV), chronopotentiometry (CV), and electrochemical impedance spectroscopy (EIS) studies. At first, the working electrode was accustomed to stabilization for continuous 20 cycles at a scan rate of 20 mV/s and then the electrode was scanned at various scan rates from 2 to 500 mV/s at a fixed potential window of −1.6 to 0.4 V (2 V). The shape of the CV curves in Fig. 3a does not representing the ideal EDLC behavior. It was noticed that there were small redox potentials indicating the formation of oxygen moieties on the surface of activated carbon. However, the shape of the CV curves remained consistent when the scan rate was increased from 2 to 500 mV/s implying good rate capability.

The following equation was used to determine the specific capacitance (Cs) values.

$$Cs = \frac{1}{vm(Vi - Vf)} \int_{Vi}^{Vf} i \times Vdv \quad \text{(F/g)}$$

where *v* is the scan rate in V/s, *m* is the active mass of the material in g, *V* is the operating potential window (*Vi* to *Vf*) in V and i is the cathodic/anodic current in A. Cyclic voltammetry studies revealed the highest capacitance of 326 F/g at 2 mV/s. The capacitance values at scan rates of 2, 5, 10, 20, 30, 40, 50, 60, 70, 80, 90, 100, 200, 300, 400 and 500 mV/s were found to be 326, 237, 191, 159, 142, 131, 122, 116, 111, 108, 103, 99, 93, 88, 80, 74, 70, 66 and 56 F/g respectively. To study the charge and discharge behavior, the electrode was subjected to chronopotentiometry studies by supplying different current densities of 1, 2.5, 5, 7.5, 10, 12.5, 15, 17.5, 20, 22.5, 25, and 30 A/g. The specific capacitance (Cs) values were calculated using the equation given below.

$$Cs = \frac{i}{m(\frac{\Delta V}{\Delta t})} \quad \text{(F/g)}$$

where *i* is the applied current density in A/g, ΔV is the operating potential window in V, Δt is the discharge time in s and *m* is the active mass of the material in g. The corresponding capacitance values were found to be 96, 58, 48, 42, 38, 35, 33, 30, 26, 22, 17, and 12 F/g. The peaked triangular-shaped curves show the EDLC curve as in Fig. 3b.

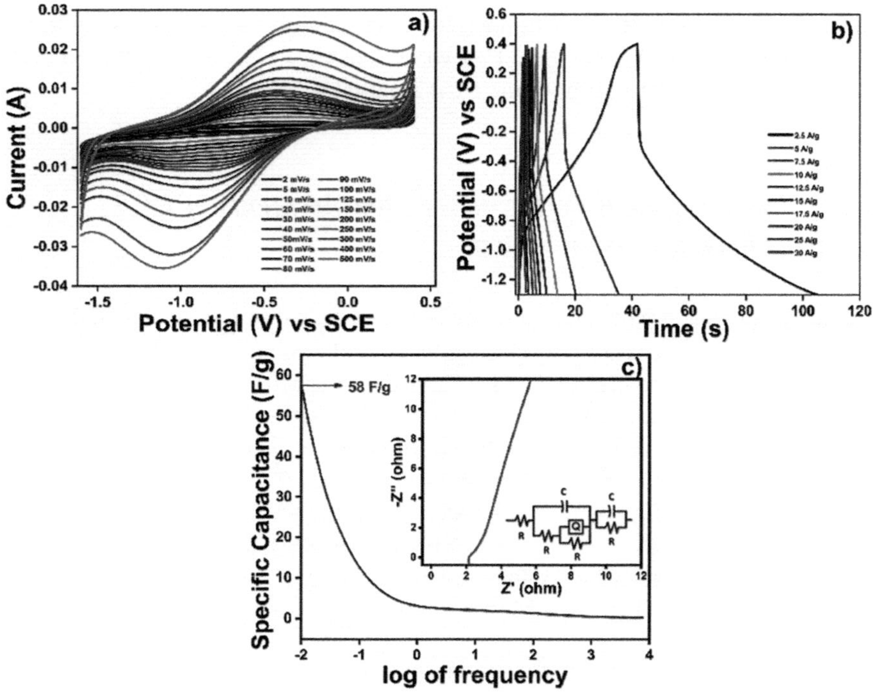

Fig. 3 **a** Cyclic voltammogram of KCB at various scan rates under three-electrode system, **b** chronopotentiometry curves of KCB at different current densities, and **c** graph of specific capacitance versus log of frequency and inset shows Nyquist plot of KCB and the equivalent circuit

The electrochemical impedance spectroscopy (EIS) studies were performed at open circuit potential in the frequency range 10^{-2}–10^{5} Hz and the specific capacitance was estimated using the following equation:

$$Cs = \frac{1}{2\pi f Z''} \quad (\text{F/g})$$

where f is the frequency, Z'' is the imaginary axis and the Cs value was found to be 58 F/g (Fig. 3c).

Prior to the activation of the carbon, the non-activated carbon was also tested for its electrochemical performance in the same manner as it was done for the KCB material. CV measurements were conducted in the potential window of −1.2 to 0.4 V (1.6 V) and the capacitance at a scan rate of 2 mV/s was found to be only 82 F/g. Similarly, the capacitance value in CP was merely 21 F/g.

3.3 *Electrochemical Studies of Symmetric Supercapacitor Device (Symmetric Coin Cell)*

To determine the operating potential window, the constructed symmetric coin cell was subjected to different operating potentials ranging from 0 to 1.2, 1.4, 1.6, 1.8, and 2 V in 1 M KOH electrolyte as shown in Fig. 4a. This experiment indicated the stability of the coin cell in the operating potential window of 0–2 V. Henceforth the coin cell was subjected to stabilization for continuous 20 cycles at 30 mV/s in the operating potential window of 0–2 V. Cyclic voltammetry studies were conducted from scan rates 2–500 mV/s to check for stability and good rate capability (Fig. 4b). The capacitance values at scan rates of 2, 5, 10, 20, 30, 40, 50, 60, 70, 80, 90, 100, 200, 300, 400, and 500 mV/s were found to be 98, 66, 51, 41, 35, 31, 29, 27, 25, 24, 22, 21, 15, 11, 9, 8 F/g, respectively. At 2 mV/s the coin cell provided the highest specific capacitance of 98 F/g. Chronopotentiometry studies revealed the working potential of the coin cell to be from 0 to 1.8 V and the trials were carried out from low to high current densities in the range 1–30 A/g as shown in Fig. 4c. Capacitance values at current densities of 1, 2.5, 5, 7.5, 10, 12.5, 15, 17.5, 20, 22.5, 25, and 30 A/g were 29, 21, 16, 13, 11, 9, 8, 7, 6, 5, 4, and 3 F/g, respectively. At 1 A/g, the coin cell provided the highest specific capacitance of 29 F/g. EIS was carried out at open circuit potential in the frequency range of 10^{-2}–10^{5} Hz to evaluate the specific capacitance and resistance of the symmetric coin cell. The specific capacitance was found to be 22.5 F/g with a solution resistance (Rs) of 1.28 Ω and charge transfer resistance (Rct) of 26 Ω (Fig. 4d).

Further tests were conducted on the same symmetric coin cell to examine the cycle stability by subjecting the coin cell to 25,000 continuous charge–discharge cycles and the calculations of coulombic efficiency, specific energy, and specific power were made using the below equations:

$$\text{Coulombic effeciency, } \eta = \frac{td}{tc} \times 100 \quad (\text{in } \%)$$

$$\text{Specific Energy, } E = \frac{Cs \times \Delta V^2}{2 \times 3.6} \quad (\text{in Wh/kg})$$

$$\text{Specific Power, } P = \frac{E}{td} \times 3600 \quad (\text{in W/kg})$$

where t_d and t_c are the discharging and charging time in s, Cs is the specific capacitance in F/g, ΔV is the operating potential window in V and E is the specific energy in Wh/kg. From the cycle stability tests, it was found that the device retained 100% coulombic efficiency and 56% capacitance retention after 25,000 cycles at a current density of 20 A/g as shown in Fig. 5a. A Ragone plot (Fig. 5b) of the designed coin cell demonstrated a specific energy of 12.9 Wh/kg with a corresponding specific power of 149.9 W/kg at a current density of 1 A/g. For the application of this KOH-activated carbon derived from banana stem (KCB), two such coin cells with an operating

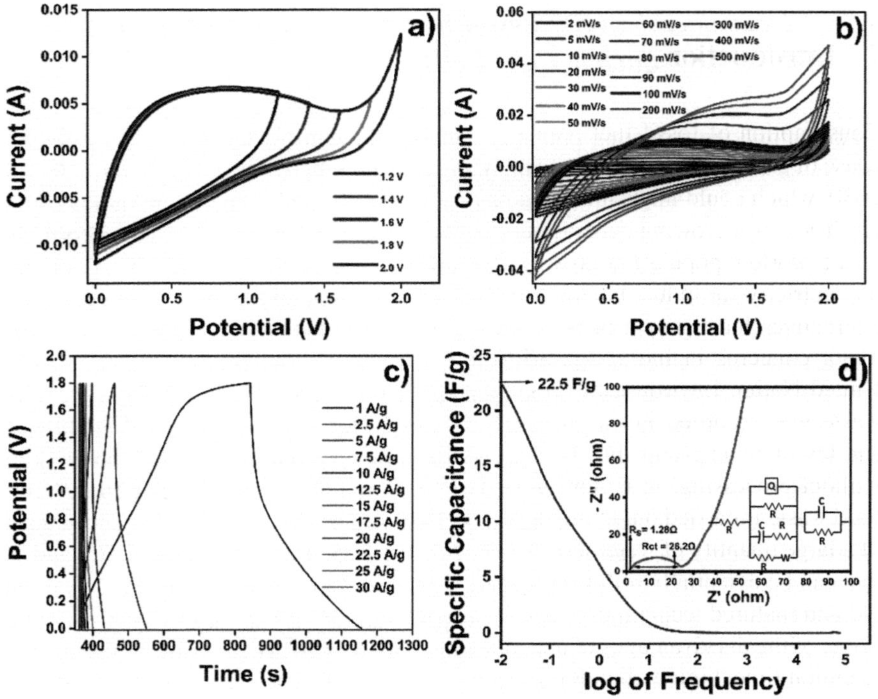

Fig. 4 **a** CV curves of a symmetric supercapacitor coin cell using KCB in different operating potential windows, **b** CV curves of coin cell at different scan rates from 2 to 500 mV/s, **c** charge discharge curves of a symmetric supercapacitor coin cell at various current densities from 1 to 30 A/g and **d** a plot of Cs vs. log of frequency from EIS data, with Nyquist graph and equivalent circuit as inset

potential range of 0–2 V were connected in series and employed to illuminate a 1-W white LED for 2 min (Fig. 5c).

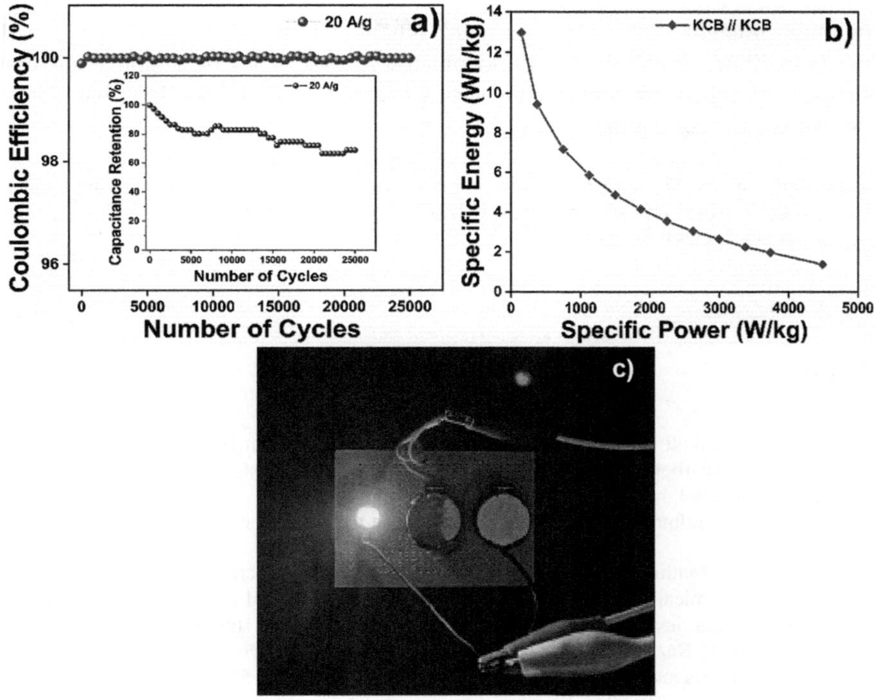

Fig. 5 **a** Plot of coulombic efficiency at corresponding cycles and inset shows the number of cycles versus capacitance retention of the symmetric coin cell, **b** Ragone plot of symmetric supercapacitor, and **c** a digital photograph of 2 coin cells connected in series used to power a 1-W LED

4 Conclusion

In this work, abundantly produced agricultural waste, i.e., stems of *Musa acuminata* (banana) were carbonized and chemically activated using KOH to synthesize porous carbon for supercapacitor application. The chemical activation was carried out by following the optimized weight ratio protocol. Firstly, the obtained activated carbon (KCB) was characterized using P-XRD, BET, SEM, and TEM techniques. The X-ray diffraction peaks and (hkl) planes confirmed the presence of hexagonal carbon. The SEM and TEM images with BET surface area of 3059 m^2/g and average pore diameter value of 1.85 nm further substantiated the formation of porous carbon. Secondly, the electrochemical performance of the KCB material was studied in three-electrode system, where the material showed a specific capacitance of 326, 96, and 58 F/g at a scan rate of 2 mV/s, current density of 1 A/g and at open circuit potential in CV, CP and EIS techniques respectively. Thirdly, a symmetric supercapacitor (coin cell CR-2032) was constructed using the KCB material and it demonstrated a high specific capacitance of 98 F/g at a scan rate of 2 mV/s and a Cs of 29 F/g at a current density of 1 A/g. Cycle stability tests were conducted on the coin cell and it exhibited 100%

coulombic efficiency and 56% capacitance retention after 25,000 cycles at a current density of 20 A/g. Finally, to test the symmetric coin cell's practical application, two of these coin cells were connected in series, charged, and were used to illuminate a 1-W white LED for 2 min.

Acknowledgements The authors thank Dr. Krishna Venkatesh, Director of the Centre for Incubation, Innovation, Research and Consultancy (CIIRC) for providing all the instrumentation facilities to carry out this research work.

References

1. Tatrari G, Karakoti M, Tewari C, Pandey S, Bohra B, Dandapat A, Sahoo N (2021) Solid waste-derived carbon nanomaterials for supercapacitor applications: a recent overview. Mater Adv 2:1454–1484. https://doi.org/10.1039/d0ma00871k
2. Food and Agricultural Organization (2022) Banana market review–preliminary results 2022 report, Rome
3. Thenappan M, Mathiyalagan K, Abdollahifar M, Rengapillai S, Marimuthu S (2023) Structural and electrochemical properties of Musa acuminata fiber derived hard carbon as anodes of sodium-ion batteries. Energies 16:979. https://doi.org/10.3390/en16020979
4. Raza W, Ali F, Raza N, Luo Y, Kim K, Yang J, Kumar S, Mehmood A, Kwon E (2018) Recent advancements in supercapacitor technology. Nano Energy 52:441–473. https://doi.org/10.1016/j.nanoen.2018.08.013
5. Liu Y, Zhao D, Liu H, Umar A, Wu X (2019) High performance hybrid supercapacitor based on hierarchical MoS_2/Ni_3S_2 metal chalcogenide. Chin Chem Lett 30:1105–1110. https://doi.org/10.1016/j.cclet.2018.12.024
6. Winter M, Brodd RJ (2004) What are batteries, fuel cells, and supercapacitors? Chem Rev 104:4245–4270. https://doi.org/10.1021/cr020730k
7. Yan J, Li S, Lan B, Wu Y, Lee P (2019) Rational design of nanostructured electrode materials toward multifunctional supercapacitors. Adv Funct Mater 30:2. https://doi.org/10.1002/adfm.201902564
8. Pandolfo AG, Hollenkamp AF (2006) Carbon properties and their role in supercapacitors. J Power Sources 157:11–27. https://doi.org/10.1016/j.jpowsour.2006.02.065
9. Kasimalla VK, Srinivasulu NG, Velisala V (2018) A review on energy allocation of fuel cell/battery/ultracapacitor for hybrid electric vehicles. Int J Energy Res 42:4263–4283. https://doi.org/10.1002/er.4166
10. Wang J, Kaskel S (2012) KOH activation of carbon-based materials for energy storage. J Mater Chem 22:23710. https://doi.org/10.1039/c2jm34066f
11. Shaker M, Ghazvini AA, Cao W, Riahifar R, Qi GE (2021) Biomass-derived porous carbons as supercapacitor electrodes—a review. New Carbon Mater 36:546–572. https://doi.org/10.1016/s1872-5805(21)60038-0
12. Taer E, Taslim R, Mustika WS, Kurniasih B, Agustino, Afrianda A, Apriwandi (2018) Production of an activated carbon from a banana stem and its application as electrode materials for supercapacitors. Int J Electrochem Sci 13:8428–8439. https://doi.org/10.20964/2018.09.55
13. Taer E, Afdal Yusra D, Amri A, Awitdrus TR, Apriwandi A, Putri A (2021) The synthesis of activated carbon made from banana stem fibers as the supercapacitor electrodes. Mater Today: Proc 44:3346–3349. https://doi.org/10.1016/j.matpr.2020.11.645
14. Dulyaseree P, Sama H, Sada S, Ukkakimapan P, Yordsri V, Sattayarut V, Wongwiriyapan W (2022) Effect of carbonization temperature on physical properties and specific capacitance of activated carbon derived from banana stem and its application as supercapacitor electrodes. Curr Appl Sci Technol 23:2. https://doi.org/10.55003/cast.2022.02.23.002

15. Mestre AS, Freire C, Pires J, Carvalhoa A, Pinto L (2014) High performance microspherical activated carbons for methane storage and landfill gas or biogas upgrade. J Mater Chem A 2:15337–15344. https://doi.org/10.1039/c4ta03242j
16. Byatarayappa G, Guna V, Venkatesh K, Reddy N, Nagaraju N, Nagaraju K (2021) Superior cycle stability performance of a symmetric coin cell fabricated using KOH activated bio-char derived from agricultural waste—Cajanus cajan stems. J Environ Chem Eng 9:106525. https://doi.org/10.1016/j.jece.2021.106525
17. Yahaya M, Ramli I, Muhamad EN, Ishak NS, Umar UI, Yap YH (2020) K_2O doped dolomite as heterogeneous catalyst for fatty acid methyl ester production from palm oil. Catalysts 10:791. https://doi.org/10.3390/catal10070791
18. Byatarayappa G, Radhika MG, Bhatta UM, Ingale D, Venkatesh K, Nagaraju N, Nagaraju K (2021) A correlation of surface/textural properties of carbon derived from Musa acuminata stem and its high supercapacitance performance with superior cycle stability. J Energy Storage 44:103424. https://doi.org/10.1016/j.est.2021.103424
19. Radhika MG, Gopalakrishna B, Chaitra K, Bhatta LKG, Venkatesh K, Kamath MK, Kathyayini N (2020) Electrochemical studies on Ni, Co and Ni/Co-MOFs for high-performance hybrid supercapacitors. Mater Res Exp 7:054003. https://doi.org/10.1088/2053-1591/ab8d5d

Recycling of Graphite Rod from Primary Batteries to Make Graphene Oxide by Electrochemical Exfoliation Method for Supercapacitor Application

Saurabh S. Suranshe, Awanikumar Patil, and Jitendra Chavhan

1 Introduction

Primary battery recycling around the world mainly focuses on recovering Zn, Mn, Fe, and other valuable metals, while graphite rods from these batteries are typically dumped in landfills [1]. This graphite rod can be used as an excellent precursor for producing different carbon-based materials. The recycling of this graphite rod guarantees a cheap and abundantly available graphite without incurring any production costs while recycling of these rods ensures a reduction in carbon footprints. Carbon-based material typically displays electric double-layer capacitor (EDLC) type behavior when used for supercapacitor application. Low production cost, longevity, and suitability with most electrolytes makes these carbon-based material very attractive for commercial energy storage applications. Activated carbon (AC) is the most popular carbon-based material owing to its high-surface area, porous structure, abundance, and low cost. However, when used for supercapacitor applications, most of the carbon atoms are not accessible to electrolytes, making it inefficient for the application. Another issue is AC's limited electrical conductivity [2].

The commercial growth of portable electronics, electric vehicles, and off-grid charge storage applications is generating demand for better and cheaper EDLC-type material. In this pursuit, graphene is a promising alternative, but producing high-quality graphene involves a large amount of resources and energy, which eventually increases production costs. On the other hand, graphene oxide (GO), the oxidized form of graphene, is relatively simple to produce, inexpensive, and provides better cyclic stability and good performance [3–14].

Traditionally, GO is synthesized by chemical oxidation of graphite using the Hummer's and Staundenmair methods. GO synthesized from these methods is

S. S. Suranshe · A. Patil (✉) · J. Chavhan
Department of Metallurgical and Materials Engineering, Visvesvaraya National Institute of Technology, Nagpur 440010, India
e-mail: appatil@mme.vnit.ac.in

S. T. Aruna et al. (eds.), *Recent Advances in Electrochemical Science and Technology*, Springer Proceedings in Materials 47, https://doi.org/10.1007/978-981-97-3958-5_4

highly decorated with oxygen-containing functional groups and has disrupted the sp^2 network. The oxygen-containing groups form covalent bonds with carbon atoms and break the continuity of the 2D structure, which reduces charge concentration and carrier mobility. These functional groups also act as the scattering center for the flow of electrons, thus reducing the overall conductivity of the material. In recent times electrochemical exfoliation (ECE) route to produce GO has been widely investigated. ECE involves low cost, simple equipment, and relatively less chemical consumption which ensures less oxidation and lower disruption in the 2D network. The basic principle of ECE involves the intercalation of ionic species into the graphite precursor under the effect of opposite charge; causing the graphene sheets to expand and subsequently exfoliate. Many precursors, including expensive highly oriented pyrolytic graphite, natural graphite and affordable graphite powder, graphite rod, and graphitic film have been used to synthesize GO using this approach. The use of central graphite rods from primary batteries has received relatively less attention as a precursor for GO synthesis and the stability of this GO for supercapcitor applications has not investigated so far.

Current work is an attempt to recycle graphite rods from exhausted primary batteries and use them as a precursor for GO synthesis by the ECE method. Cyclic voltammetry (CV), galvanostatic charge–discharge (GCD), and electrochemical impedance spectroscopy (EIS) are used to explore the electrochemical and supercapacitive properties of the exfoliated GO. The longevity of GO-electrodes is also evaluated using 5000 cycles of stability testing.

2 Experimental

Graphite rod for exfoliation was salvaged from a C-type primary battery using the ECE method [6]. H_2SO_4 (99.99% pure), N-Methyl-2-Pyrrolidone (NMP, 99.13% pure), and ethanol were purchased from fisher scientific. Poly(vinylidene-fluoride) (PVDF) was purchased from Ottokemi Pvt Ltd, India, and carbon black from Loba Chemicals. Grade 316-SS (Gauge 46) sheet was purchased from the local market.

ECE was carried out in a two-electrode electrochemical cell with the graphite rod as the working electrode and the platinum rod as a counter electrode, in 0.1M H_2SO_4 electrolyte. For the first 10 min, the voltage was set to 2 V, latter on the voltage was raised to 10 V and was maintained for the next 5 h. The exfoliated material was filtered using filter paper. After washing with DI water and rinsing with ethanol in an ultrasonic bath, GO was oven-dried for 12 h at 80 °C. For electrode preparation, the SS-316 substrate was polished to remove any oxides using sandpaper with grit size from 1000 to 2000. In 0.5 mL NMP, a mixture of 80% GO, 10% carbon black, and 10% PVDF was prepared. This mixture was painted over a polished SS-316 substrate and dried for 12 h at 80 °C. These sheets were cut into 40 mm $\times$ 10 mm after drying for electrochemical examination. Details of reagents used and equipment for physical and chemical characterization are discussed in supplementary data.

Field Emission Scanning Electron Microscopy (FE-SEM) by FE-SEM, JEOL-JSM7610F, and High-Resolution Transition Electron Microscopy (HR-TEM) by FEI Tecnai G2, F30 were used to study the surface morphology of exfoliated GO powder. Structural information was studied with XRD (X'Pert Pro Multi-Purpose Diffractometer in the 2θ range of 10°–60°). Raman spectra were collected with a Horiba Scientific Spectrometer in the range of 1000–3000 cm^{-1}. Thermogravimetric analysis (TGA) by TA SDT600 thermal analyzer, surface area measurement using Brunauer-Emmet-Teller (BET) method by N_2 absorption–desorption method using NOVA touch LX4 surface and pore size analyzer.

The electrochemical and supercapacitor properties of GO were investigated by CV in the voltage range of –0.3 to 0.4 V, Galvanostatic charge–discharge (GCD) between 1 and 2 mA cm^{-2} and electrochemical impedance spectroscopy (EIS) between 0.1 Hz and 100 kHz in 1M H_2SO_4 solution using Biologic VMP-300 workstation.

Constant power (Cp) and specific capacitance were calculated using the following equations:

$$Cp = \frac{1}{m(v1 - v2)k} \int_{v1}^{v2} I(v)dv \tag{1}$$

Cp is specific capacitance, *m* is an active material mass on the electrode, $(v1 - v2)$ voltage range, and *I* is the current density.

$$Cs = \frac{I \times \Delta T}{(v2 - v1) \times m} \tag{2}$$

Cs is specific capacitance, *I* is charge–discharge current, $(v2 - v1)$ Potential window, *m* is working electrode mass, and ΔT is discharge time.

3 Result and Discussion

During a few minutes of exfoliation, a trail of dark particles was observed leading toward a positive platinum electrode. This movement of exfoliated particles could be due to the overall negative charge on the particles due to oxidation. Further, in the experiment electrolyte turns dark black unlike oxidation by chemical oxidation process where the electrolyte solution turns yellow to brown [1, 2]. This suggests lesser oxidative attack by electrochemical exfoliation compared to the chemical process. This also signifies that the GO produced by the ECE process retains metallic structure better than that produced by the chemical oxidation process. Weight of graphite rod after exfoliation was 3.7 g which accounts for 50% exfoliation in 5 h.

3.1 Physical Characterization

Figure 1a shows the XRD pattern of the precursor graphite rod and GO. Both the patterns show typical graphitic peaks at $2\theta = 26°$ and $43°$ corresponding to (002) and (101) planes respectively [9]. The XRD spectra for GO show significantly reduced intensity with an increase in width for the peak at 26° signifying reduced crystallinity and increased defects in the sp^2 network. The XRD pattern of GO shows a small hump around $2\theta = 12°$ with a d-spacing of 0.752 nm (from Braggs Law). The increase in d-spacing (from 0.34 to 0.752 nm) and hump around 12° is due to the combined effect of oxidation of precursor graphite during the exfoliation process, disruption of AB structure, and intercalation of H_2O molecule [14].

TGA was used to test the thermal stability of GO, where GO was heated from room temperature to 800 °C in a nitrogen environment. Weight loss was observed in three stages and is in agreement with the reported literature [3, 4] with a total weight loss of 43.13%. upto 800 °C. Figure 1b shows the TGA profile for GO. The first stage of weight loss for GO begins below 100 °C and continues until 130 °C; this weight loss is due to evaporation of absorbed moisture and accounts for 5–6% of total weight loss [5, 6]. Weight loss in the second stage starts at 130 °C, peaks

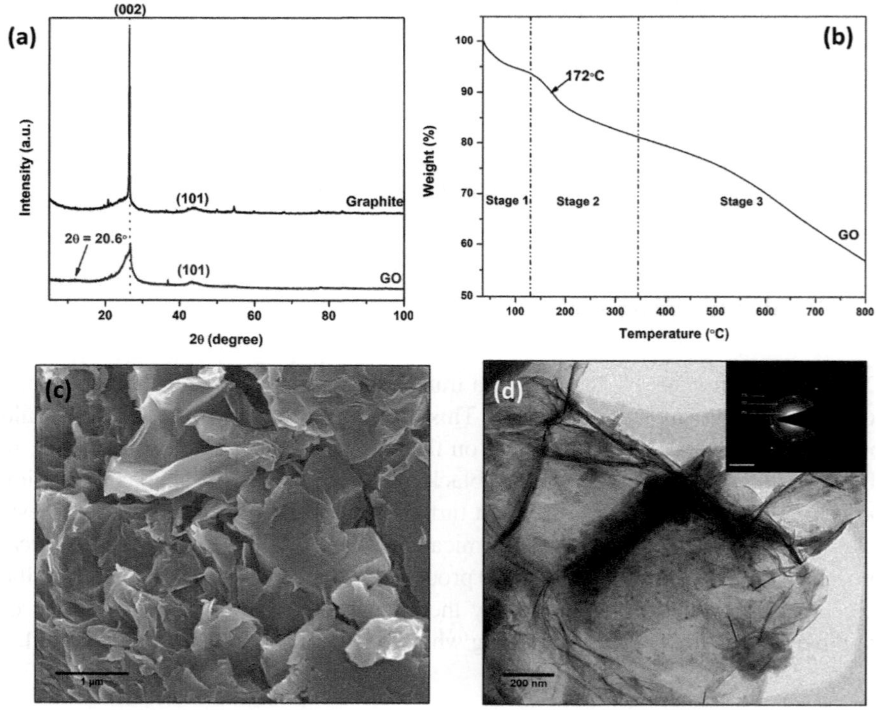

Fig. 1 **a** XRD spectra of precursor graphite rod and exfoliated GO, **b** TGA analysis of GO, **c** FE-SEM image of GO showing sheet-like morphology, **d** HR-TEM images of GO sheets

at 174 °C, and lasts until 348 °C. This weight loss is attributed to the pyrolysis of oxygen-containing functional groups and is 13–14% of total weight loss [3, 4]. The last stage ranges from 350 to 800 °C and is because of the combustion of the carbon skeleton of graphene [3, 6]. In the last stage, the weight of GO was observed to have reduced by 25% and is the maximum of all stages signifying the presence of a sufficient number of disconnected carbon atoms.

FE-SEM images of GO show a thin crumbled 2D paper-like structure (Fig. 1c) with no charging effect, signifying GO papers are electrically conductive. GO sheets feature curled and abruptly broken edges, which could be due to pressures induced by oxidation and intercalation of ionic species. Despite their thinness, the sheets are sufficiently flexible, as evidenced by various bends in the sheets that did not fracture. Another noticeable feature of the sheets is the different C/O ratios at different sites. EDS studies of GO show higher oxygen content at the edge site relative to the longitudinal surface (C/O ratio on edge sites by EDS is 5 and for the longitudinal surface is 16), this is due to the higher concentration of oxygen functionalities on edge sites relative to longitudinal plane. HR-TEM images of partly transparent GO sheets are shown in Fig. 1d. Light regions represent few layers of GO, while gray and black portions imply bundled GO sheets. The SAED (selective area electron diffraction) pattern provided as the inset in the HR-TEM image shows individual spots that are attributed to the polycrystalline nature of GO sheets. This SAED pattern validates the presence of (002), (101), and (110) planes which also confirms with XRD spectra shown in Fig. 1a.

XPS analysis was used to study the chemical nature of GO. Wide scan XPS spectra of GO in Fig. 2a, show the presence of carbon and oxygen elements only. The intensity of the oxygen peak in GO spectra is comparable to that of the carbon peak suggesting high oxidation on the graphene plane. The deconvoluted C1s peak of GO (Fig. 2b) has sp^2 hybridized C=C(72.72%) peak at 284.66 eV with a significant presence of C–O (27.29%) at 286.63 eV and some traces of O–C=O at 289.28 eV [11]. Specific surface area and pore size distribution of GO were analyzed using N_2 absorption–desorption by BET analysis. Figure 2c shows N_2 absorption–desorption graphs for GO, respectively. The GO surface area analysis indicates 18.47 $m^2\ gm^{-1}$ and 0.046 cc gm^{-1}.

Raman spectroscopy is an important characterization technique related to carbon-based materials. A typical graphitic Raman spectrum shows a G peak around 1575 cm^{-1}, caused by optical E2g phonons in the Brillouin Zone and bond stretching of sp^2 atoms in chains and rings. The second peak in graphitic material is approximately at 1355 cm^{-1} [8] this peak is due to the defects in the carbon plane, such as sp^2 network discontinuities, the presence of secondary bonds, point defects, grain boundaries, etc. [12]. The third peak, G′ or 2D, is around 2700 cm^{-1} and is linked to second-order zone phonons [13]. Raman spectra of GO presented in Fig. 2d shows a G peak at 1592 cm^{-1}, a D peak at 1338 cm^{-1}, and a 2D peak at 2706 cm^{-1}. The D peak in GO has a comparable intensity to the intensity of the G peak with an I_D/I_G ratio (intensity ratio of D and G peaks) of 0.95. The strong D peak in GO suggests a significant disorder in structure brought upon by ECE.

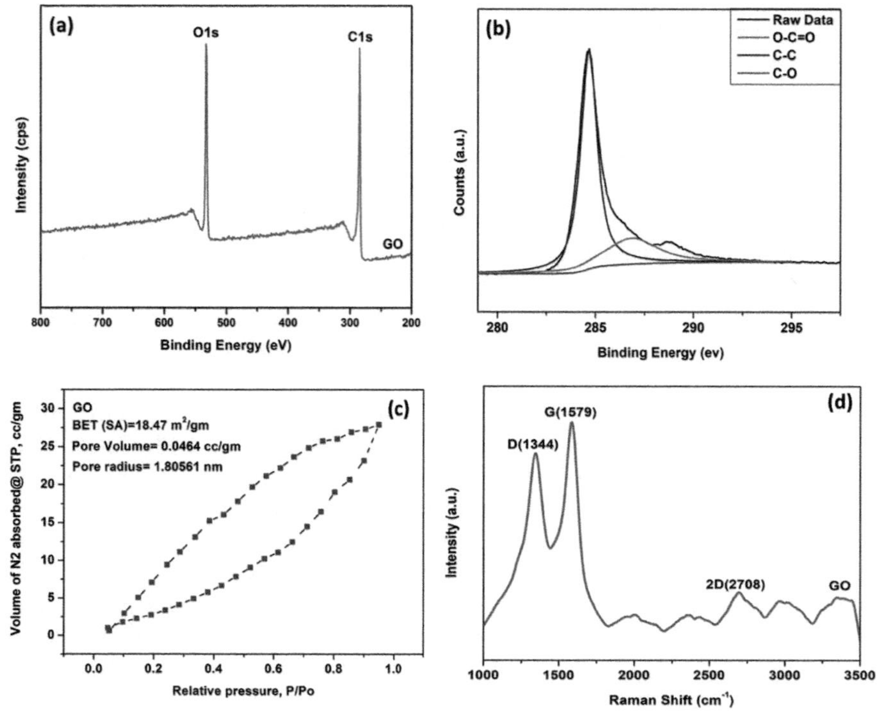

Fig. 2 **a** XPS full scan spectra of exfoliated GO, **b** high-resolution spectra of C1s region of GO, **c** N_2 absorption–desorption isothermal plot of GO, **d** Raman spectra of GO

3.2 Electrochemical Studies

Supercapacitive properties of GO electrode were explored using CV in the potential range of −0.3 to 0.4 V in 1M H_2SO_4 aqueous electrolyte. Figure 3a shows the quasi-rectangular CV curve of the GO electrode at all scan rates (5–200 mV s^{-1}) indicating typical EDLC behavior. XPS plots show the presence of oxygen fictional groups introduced after exfoliation. The presence of these oxygenated functional groups reduces the total conductivity of the GO and is responsible for the effect of leakage potential seen at the ends of the CV curve [14]. The specific capacitance of the electrode was calculated from CV using Eq. 1 in supplementary data. The CV calculations for the GO electrode show a maximum capacitance of 39.39 F g^{-1} at 5 mV s^{-1} scan rate. Figure 3b shows the GO electrode's GCD measurement from 2 to 0.8 mA. A constant IR drop of 0.3 V is observed at all the current densities followed by a nearly linear drop in voltage with a constant slope which is a typical EDLC characteristic. The IR drop is attributed to the total resistance of the material. The specific capacitance is also calculated from the GCD curve according to the following equation. The maximum capacitance of the electrode is 57.7 F g^{-1} at 0.8 mA, with a retention of 53.35% at the maximum scan rate of 2 mA. The specific capacitance versus current density curve for GO is shown in Fig. 3c and the data is in Table 1.

The resistive behavior of GO was further characterized using the EIS technique within the frequency range of 0.1 Hz to 100 kHz. EIS data was analyzed using a Nyquist plot and is shown in Fig. 3d with a corresponding fitted circuit resembling Randles Circuit. The Nyquist plot shows a non-zero real impedance axis intercept

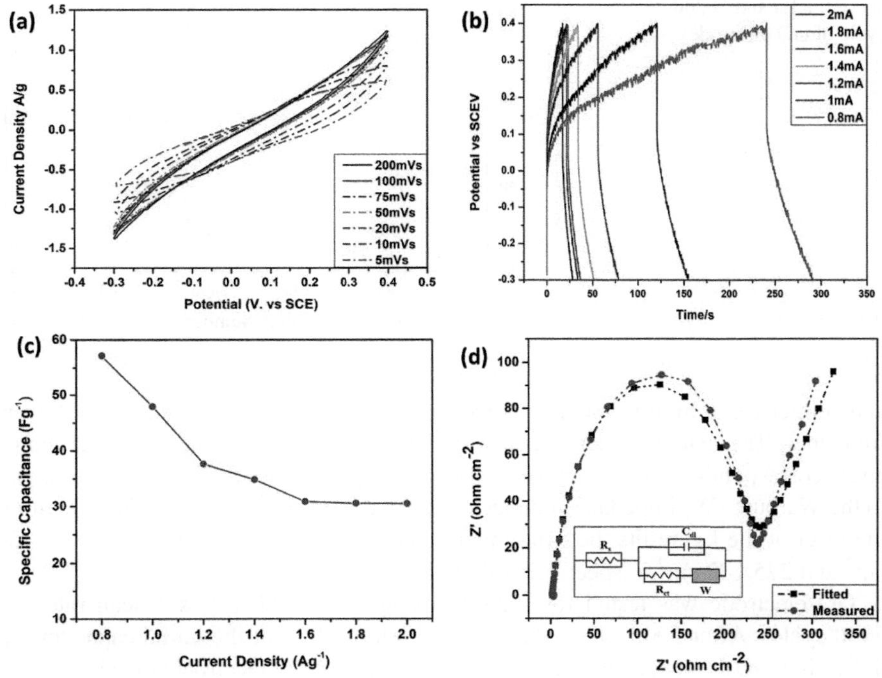

Fig. 3 **a** CV profile of GO electrode at various scan rates (5–200 mV s^{-1}), **b** GCD curves of GO electrode for various current densities (0.8–2 mA), **c** variation of specific capacitance of GO electrode versus different current densities, **d** Nyquist plot and the equivalent circuit utilized to model impedance data

Table 1 Charge discharge results of GO electrode

Specific current (mA)	Specific capacitance (F g^{-1})
0.8	**57.14**
1.0	47.95
1.2	37.68
1.4	34.84
1.6	30.85
1.8	30.53
2	30.48

Table 2 EIS equivalent circuit parameters of GO electrode

Parameters	R_s (Ω cm^2)	R_{ct} (Ω cm^2)	C_{dl} (mF cm^{-2})	W
Values	2.023	225.1	0.106	24.39

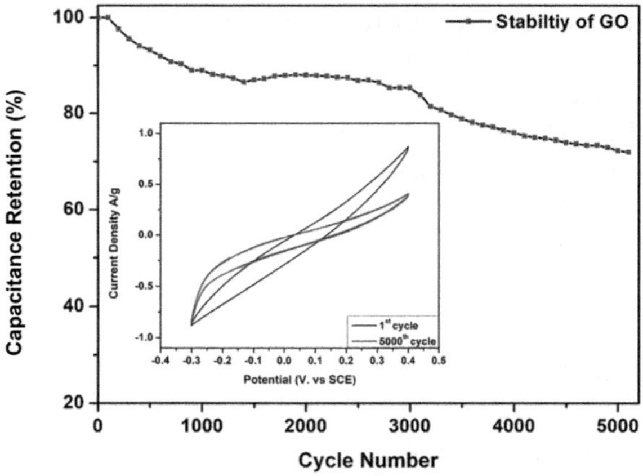

Fig. 4 Cyclic stability measurement over 5000 cycles (inset displays CV curve for 1st and 5000th cycle of GO electrode)

and semicircle at high to medium frequency followed by an inclined line at lower frequency. The real axis intercept corresponds to solution resistance (R_s), the semicircle corresponds to charge transfer resistance (R_{ct}) whereas the inclined line is due to the Warburg (W) impedance connected to conductive ion diffusion. The circuit in the inset of the figure fits the experimental data, with Rs and Rct values of 2.05 Ω cm^2 and 225.1 Ω cm^2, respectively (Table 2).

GO electrode was tested for stability using CV at 100 mV s^{-1} scan rate for 5000 cycles. After 5000 cycles, the material retains 70% of the initial capacitance, indicating GO has excellent stability. Figure 4 shows the electrochemical stability plot of the GO electrode with an inset for the 1st and 5000th CV cycle.

4 Conclusion

In this work, ECE method is effectively applied for the synthesis of GO. The ECE process proves to be simple and is easily scalable as it uses a cost-effective setup for operation. Used graphite rod from depleted dry cell batteries reduces GO production costs and carbon footprints and promotes recycling. As produced GO has sheet-like structure and good physical characteristics. The GO has a total surface area of 18.47 m^2 gm^{-1}. Electrochemical studies of GO-electrodes show good supercapacitive properties with typical EDLC-type behavior. The highest capacitance of 57.7 F g^{-1} at a charge–discharge current density of 0.8 mA was obtained for the GO electrode. Excellent stability of GO electrode has been witnessed with 70% retention of capacity at the 5000th cycle. This study shows that the depleted dry cell battery graphite rod can be effectively used as a low-cost precursor to produce GO using ECE. Based on the electrochemical investigations, synthesized GO has a strong potential for use as a very stable supercapacitor material and can effectively replace AC as a filler material.

References

1. Bandi S, Ravuri S, Peshwe DR et al (2019) Graphene from discharged dry cell battery electrodes. J Hazard Mater 366:358–369. https://doi.org/10.1016/j.jhazmat.2018.12.005
2. Kandasamy SK, Kandasamy K (2018) Recent advances in electrochemical performances of graphene composite (graphene-polyaniline/polypyrrole/activated carbon/carbon nanotube) electrode materials for supercapacitor: a review. J Inorg Organomet Polym Mater 28:559–584. https://doi.org/10.1007/s10904-018-0779-x
3. Korkmaz S, Kariper A (2020) Graphene and graphene oxide based aerogels: synthesis, characteristics and supercapacitor applications. J Energy Storage 27. https://doi.org/10.1016/j.est.2019.101038
4. Pei S, Cheng HM (2012) The reduction of graphene oxide. Carbon N Y 50:3210–3228. https://doi.org/10.1016/j.carbon.2011.11.010
5. Parvez K, Li R, Puniredd SR et al (2013) Electrochemically exfoliated graphene as solution-processable, highly conductive electrodes for organic electronics. ACS Nano 7:3598–3606. https://doi.org/10.1021/nn400576v
6. Suranshe SS, Awanikumar P (2023) Strategically improving electrical conductivity of reduced graphene oxide through a series of reduction processes. Mater Lett 333:133648. https://doi.org/10.1016/j.matlet.2022.133648
7. Liu N, Luo F, Wu H et al (2008) One-step ionic-liquid-assisted electrochemical synthesis of ionic-liquid-functionalized graphene sheets directly from graphite. Adv Funct Mater 18:1518–1525. https://doi.org/10.1002/adfm.200700797
8. Alam SN, Sharma N, Kumar L (2017) Synthesis of graphene oxide (GO) by modified hummers method and its thermal reduction to obtain reduced graphene oxide (rGO). Graphene 06:1–18. https://doi.org/10.4236/graphene.2017.61001
9. Loryuenyong V, Totepvimarn K, Eimburanapravat P et al (2013) Preparation and characterization of reduced graphene oxide sheets via water-based exfoliation and reduction methods. Adv Mater Sci Eng 2013. https://doi.org/10.1155/2013/923403
10. Zhu Y, Murali S, Stoller MD et al (2010) Microwave assisted exfoliation and reduction of graphite oxide for ultracapacitors. Carbon N Y 48:2118–2122. https://doi.org/10.1016/j.carbon.2010.02.001
11. Al-Gaashani R, Najjar A, Zakaria Y et al (2019) XPS and structural studies of high quality graphene oxide and reduced graphene oxide prepared by different chemical oxidation methods. Ceram Int 45:14439–14448. https://doi.org/10.1016/j.ceramint.2019.04.165
12. Tuinstra F, Koenig JL (1970) Raman spectrum of graphite. J Chem Phys 53:1126–1130. https://doi.org/10.1063/1.1674108
13. Nemanich RJ (1979) First and second order Raman scattering from finitesize crystals of graphite. Am Phys Soc 20:392–401. https://doi.org/10.1103/PhysRevB.20.392
14. Wang C, Zhou J, Du F (2016) Synthesis of highly reduced graphene oxide for supercapacitor. J Nanomater 4840301:1–8. https://doi.org/10.1155/2016/4840301

Facile Hydrothermal Synthesis of Boron and Nitrogen-Doped RGO as Electrode Material for Supercapacitors

Monika Shrivastav, Shubham Yadav, and Rajnish Dhiman

1 Introduction

Supercapacitors (SCs) are electrochemical devices that can store and release large amounts of charge in a short span of time. The idea of a supercapacitor was introduced by Becker in 1957, while Nippon Electric Company commercialized it for the first time as a power-saving unit in the year 1971 [1]. SCs have certain advantages like fast charging, long cycle life, low cost per cycle, high power density, high current discharge ability, etc., which makes it a potential candidate to have applications in power equipment such as voltage stabilization, power supplies and backup for large data centers, medical equipment, military equipment, micro-grids, street lights, etc. [2, 3]. Based on the charge storage mechanisms, supercapacitors can be categorized into pseudocapacitors and electrical double-layer capacitors (EDLCs). Pseudocapacitors mainly utilize transition metal oxides (such as RuO_2, TiO_2, V_2O_5, MnO_2, Co_3O_4, etc. [4]) and sulfides (like MoS_2, WS_2, etc. [5]) to generate capacitance through the Faradic redox reactions of active materials. The pseudocapacitors possess high specific capacitance and energy density but display poor electronic conductivity. EDLCs electrostatically store charge in the form of a double-layer at the electrode/electrolyte interface and are generally fabricated using carbon derivatives such as graphene, CNTs, mesoporous carbon, etc. [6]. They typically have high conductivity, remarkable durability, and long cycle life.

Graphene is a 2D material with a honeycomb lattice and hexagonally arranged sp^2-hybridized carbon framework. Owing to the high specific surface area, high thermal conductivity, and exceptional electronic conductivity, graphene has garnered significant attention for its potential use in energy storage devices [7]. It is challenging to produce graphene in large amounts, so it could be substituted by reduced graphene

M. Shrivastav · S. Yadav · R. Dhiman (✉)
Department of Physics, Malaviya National Institute of Technology, Jaipur, Rajasthan 302017, India
e-mail: rajnish.phy@mnit.ac.in

S. T. Aruna et al. (eds.), *Recent Advances in Electrochemical Science and Technology*, Springer Proceedings in Materials 47, https://doi.org/10.1007/978-981-97-3958-5_5

oxide (RGO), which has similar characteristics to graphene. RGO could be produced in higher quantities by chemically reducing graphene oxide (GO) [8]. RGO usually contains structural defects and residual oxygen-containing functional groups [9] that may result in lowered capacitance. The capacity enhancement of graphene or RGO can be achieved by doping with electron-accepting or electron-donating atoms such as phosphorous, boron, nitrogen, and sulfur [10, 11]. These heteroatoms can substitute carbon in graphene or reduce graphene oxide's layered structure. The atomic sizes of boron and nitrogen are comparable to the atomic size of carbon; thus, nitrogen and boron are the commonly used dopant heteroatoms. Boron (B) is electron deficient as it has the absence of π-electrons [12], thus, it acts as an electron acceptor. The nitrogen (N) heteroatom can also significantly alter the physical and chemical characteristics of RGO as it has two lone pairs of electrons and is more electronegative than carbon [13]. Many researchers are paying attention to doping heteroatom in RGO using various routes. Mishra et al. fabricated an RGO//N-RGO supercapacitor with a 2.2 V wide potential window [13]. Muthu et al. [11] synthesized boron-doped RGO by electrochemical exfoliation and analyzed the device performance of symmetric supercapacitor, and they reported an energy and power density of 6.45 Wh kg^{-1} and 8000 W kg^{-1}, respectively. Yadav et al. [14] reported a specific capacitance of 514 F/g for N-RGO sheets synthesized by the solvothermal method. The hydrothermal route for synthesizing B-doped and N-doped graphene has gained attention due to its high production rate and simple methodology. The hydrothermally synthesized heteroatom-doped RGO has been observed to have an order of electrochemical capacitance as N/S-RGO > N-RGO > S-RGO $\geq$ B-RGO > I-RGO [15]. In this report, graphene oxide is synthesized using a modified Hummer's method and it is reduced to form a layered structure of RGO using hydrazine. The boron and nitrogen doping of RGO is performed using boric acid and ammonia as precursors. Susanti et al. [16] reported an increase in the electric double-layer capacitance of the boron-doped RGO with an increase in the boron concentration and it was attributed to increase in electrical conductivity. Mishra et al. [17] fabricated N-doped RGO-based solid-state symmetric supercapacitor device with specific capacity of 141.1 mAh g^{-1} and energy density of 28.2 Wh kg^{-1}. The device retained 95.4% of the specific capacitance after 10,000 galvanostatic charge–discharge (GCD) cycles [17]. Most of the literature on RGO-based supercapacitor devices reports aqueous electrolytes, whereas there are a few reports on gel-polymer electrolytes-based supercapacitor devices. This work reports the capacitive performance of nitrogen and boron-doped RGO supercapacitor devices using polymer gel electrolytes. Graphene oxide (GO) has been synthesized by modified Hummer's method followed by GO reduction by hydrazine and doping by boric acid and ammonia precursors.

2 Experimental

2.1 *Materials*

Graphite flakes (99.8%), and acetylene black were purchased from Alpha Aesar. The following chemicals were procured from E-Merck, India: hydrogen peroxide (H_2O_2, 30 wt%), potassium permanganate ($KMnO_4$, 99.5 wt%), sodium nitrate ($NaNO_3$, 99 wt%), ammonia anhydrous (99.98%), and hydrazine hydrate (H_6N_2O, 99 wt%). Carboxymethyl cellulose (CMC-extra pure), and sodium borohydrate ($NaBH_4$-97%) were procured from Loba Chemie. Isopropyl alcohol (IPA-95%) and potassium hydroxide (KOH-85%) were purchased from CDH, India. All the chemicals and reagents were utilized as received, without further purification.

2.2 *Synthesis of RGO, N-RGO, and B-RGO*

Graphene oxide (GO) is synthesized using graphite flakes by the conventional modified Hummer's method [18] and is later reduced using hydrazine. In a typical procedure, 1 g G.O. is dispersed in de-ionized (DI) water (1 mg/mL) and 10 μL of hydrazine is added to the solution. The solution is then poured into a Teflon-lined autoclave which is then kept in a hydrothermal oven at 200 °C for 24 h. The solution is filtered and dried to get a black powder of reduced graphene oxide (RGO).

The boron and nitrogen doping of RGO (B-RGO & N-RGO) is obtained by adding 3 g of boric acid or 10 μL ammonia solution, respectively, to the dispersion containing 100 mg RGO in 100 mL of de-ionized (DI) water. The solution is transferred to the Teflon-lined autoclave and is then hydrothermally treated at 160 °C for 16 h. Precipitates are collected after cooling and are then washed and dried.

2.3 *Device Fabrication*

The RGO, N-RGO, and B-RGO-based supercapacitor devices were fabricated by coating an ink made of the active material over the stainless steel (SS) sheets of 1 × 1 cm^2 dimensions and 1 mm thickness. The ink was prepared by sonicating the active material, acetylene black, and PVDF solution (8:1:1) in ethanol for 2 h. The obtained slurry was coated over the electrodes and dried for 2 h at 120 °C. The SS sheets were cleaned with 1M HCl, acetone, and DI water.

The devices were fabricated using RGO as an anode and N-RGO and B-RGO as cathode materials (RGO//N-RGO and RGO//B-RGO). The negative electrodes (anode) were prepared by coating a slurry of RGO over the SS sheets and were dried. The positive electrodes (cathode) were prepared by coating a thin layer of active material (N-RGO or B-RGO) over SS sheets. The devices were assembled by sandwiching PVA (polyvinyl alcohol)-CMC (carboxymethyl cellulose)/KOH-based gel-polymer electrolytes in between positive and negative electrodes. The mentioned gel-polymer electrolyte was fabricated by soaking a gel polymer made of PVA and CMC in 3M KOH for 24 h. The procedure for the synthesis of polymer is given elsewhere [19].

2.4 Physicochemical Characterizations

The materials were characterized by X-ray diffraction (XRD), recorded by a Panalytical Xpert Pro X-ray diffractometer at a step size of 0.02°. The Raman spectroscopy was performed by WITec Micro-Raman Spectrometer using a 532 nm laser. The surface morphology of samples was investigated by a scanning electron microscope (JEOL, model JSM IT500) operated at an acceleration voltage of 10 kV. The elemental quantification of the samples was carried out by energy-dispersive X-ray spectroscopy (EDX) performed using the EDX system attached to the scanning electron microscope (SEM).

2.5 Electrochemical Characterizations

The capacitive performance of the samples was tested using an electrochemical workstation (CH-Instruments, model 760E) by cyclic voltammetry (CV) and galvanostatic discharge (GCD) in a three-electrode testing setup and GCD of the fabricated device. CV is performed in the potential window of –0.8 to 0.2 V at various scan rates, ranging from 10 to 100 mV/s in 1M KOH electrolyte solution. The GCD (in the three-electrode setup) curves are recorded at different current densities in 1M KOH. Electrochemical impedance spectroscopy (EIS) was performed in the frequency range of 0.01 Hz to 4 MHz at an amplitude of 5 mV.

The measurements were recorded by depositing the sample over a glassy carbon electrode. An ink of active material (N-RGO/B-RGO) was prepared in IPA solvent with acetylene black and carboxymethyl cellulose (CMC) as a binder by mixing in the ratio of 7:2:1; the mixture was sonicated until a homogenous slurry was obtained. The working electrode was prepared by drop-casting the prepared ink over a glassy carbon electrode of 3 mm diameter. Pt-wire was used as a counter electrode, and Ag/AgCl electrode was used as a reference electrode. 1M KOH was used as an electrolyte.

The specific capacitances (CS) of RGO, N-RGO, and B-RGO samples were calculated from CV curves using Eq. (1) [5].

$$C_s = \frac{A}{m \times k \times \Delta v} \tag{1}$$

where A is the area enclosed under the CV curve ($A = \int I.dt$), m is the total mass loading on electrode (g), k is the scan rate (mV/s), and Δv is the range of potential window (1 V). The mass loading on the glassy carbon electrode is 150 μg.

Equation (2) has been used to calculate the specific capacitance from GCD curves at different current densities [4].

$$C_S = \frac{I \times \Delta t}{m \times \Delta V} \tag{2}$$

where I/m is the current density (A/g), Δt is the discharge time of a cycle, m is the mass loading on the working electrode, and ΔV is the potential window [6]. The specific capacitance has been calculated at current densities of 1, 2, 3, and 4 A/g.

The specific capacitance of the RGO//N-RGO and RGO//B-RGO devices is also calculated using Eq. (2). The energy density (E) in units of Wh/kg and power density (P) in units of W/kg is calculated using Eq. (3) and dividing energy density by discharge time (Δt) is calculated using Eq. (4).

$$E = \frac{C_S \times \Delta V^2}{2}. \tag{3}$$

$$P = \frac{E}{\Delta t} \tag{4}$$

The terms ΔV and Δt are potential range and discharge time (defined above) [19].

3 Results and Discussion

3.1 Physicochemical Characterization

X-ray diffraction is used to analyze the crystallinity of the samples (Fig. 1a). XRD pattern of all these samples shows a characteristic peak at 26° corresponding to (002) reflection plane of the reduced graphene oxide. Another low-intensity peak appears around 2θ = 43.5°, this can be indexed to (004) reflection plane of natural graphite [20]. Some shoulder peaks comprise a hump, possibly due to the exfoliated form of GO sheets [21]. The average crystallite size of RGO, N-RGO, and B-RGO were calculated by the Debye–Scherrer formula [22] by considering the peak corresponding to (002) reflection. The values of crystallite size were calculated using Gaussian fit on XRD data. The calculated values of crystallite size of RGO, B-RGO, and N-RGO are 1.9, 2.0, and 2.4 nm, respectively. The increase in the crystallite size value suggests a slight increase in crystallinity upon doping.

Figure 1b shows the Raman spectra of RGO, N-RGO, and B-RGO. All three samples exhibit two peaks corresponding to the D and G bands. The D band appears at 1349, 1341, and 1338 cm^{-1} for RGO, N-RGO, and B-RGO, respectively, whereas

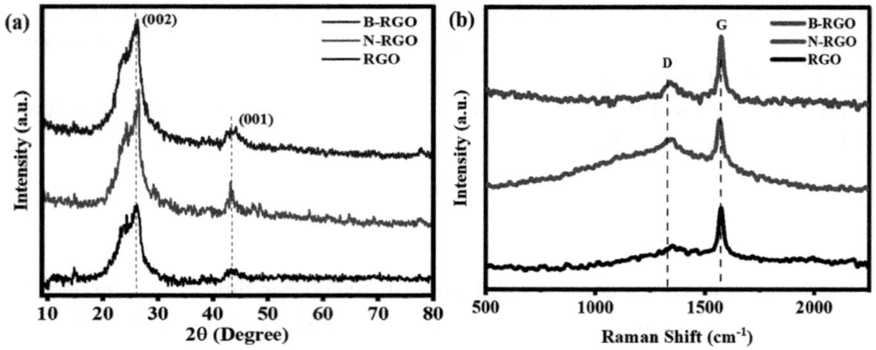

Fig. 1 **a** XRD and **b** Raman data of RGO, N-RGO, and B-RGO

the G-band appears at (i) 1572 cm^{-1} for RGO and B-RGO, and (ii) 1566 cm^{-1} for N-RGO. The G-band occurs due to the in-plane stretching of C–C in sp^2-hybridized graphitic rings, whereas the D band appears due to the presence of defects in the graphitic planes that can also be considered as the presence of sp^3-hybridized carbon. The D band is blue-shifted in N and B-doped RGO, while the G-band is red-shifted in B-doped RGO. The red shifting of the G-band arises due to the restacking of layers [23] upon doping. The calculated value of the disorder parameter (I_D/I_G) for RGO, B-RGO, and N-RGO are 0.27, 0.37, and 0.51, respectively; this suggests that doping in RGO leads to the creation of defects in the graphitic planes of RGO [24].

The morphology of the samples is analyzed by SEM (Fig. 2); all the samples appear to have a sheet-like structure. A comparison of SEM images indicates the smaller particle size of N-RGO (Fig. 2c, d) and B-RGO (Fig. 2e, f) than RGO (Fig. 2a, b). Slight differences are observed in the morphology of RGO, N-RGO, and B-RGO; this happens due to the restacking of layered graphene sheets caused by the doping of heteroatoms (nitrogen and boron).

The elemental composition of the samples has been calculated by EDX spectra (Fig. 3) and the atomic% is given in the table in the inset images. The EDX spectra of RGO only show C and O in 73% and 27% atomic concentrations, respectively. The spectra of N-RGO and B-RGO show a significant amount of nitrogen (36%) and boron (35%) in the respective samples. There is a drastic decrease in the amount of oxygen in both samples upon doping; this demonstrates a high level of doping in the RGO. A small peak at 3.6 keV, corresponding to calcium [25] has been observed in both the treated samples, possibly due to some contamination during the hydrothermal treatment.

Fig. 2 SEM images of **a**–**b** RGO, **c**–**d** N-RGO, and **e**–**f** B-RGO

3.2 Electrochemical Characterization

3.2.1 CV and GCD

CV was performed in 1M KOH electrolyte to examine the capacitive performance of the synthesized materials as a supercapacitor electrode. The CV profile of the samples (Fig. 4a) appears to be quasi-rectangular with significant double-layer capacity. The quasi-rectangular shape suggests a diffusive mechanism. A slight pseudocapacitive behavior is also present at the edges of CV profiles of RGO, N-RGO, and B-RGO due to redox reactions of electrolyte ions with electrode functional groups [23]. The

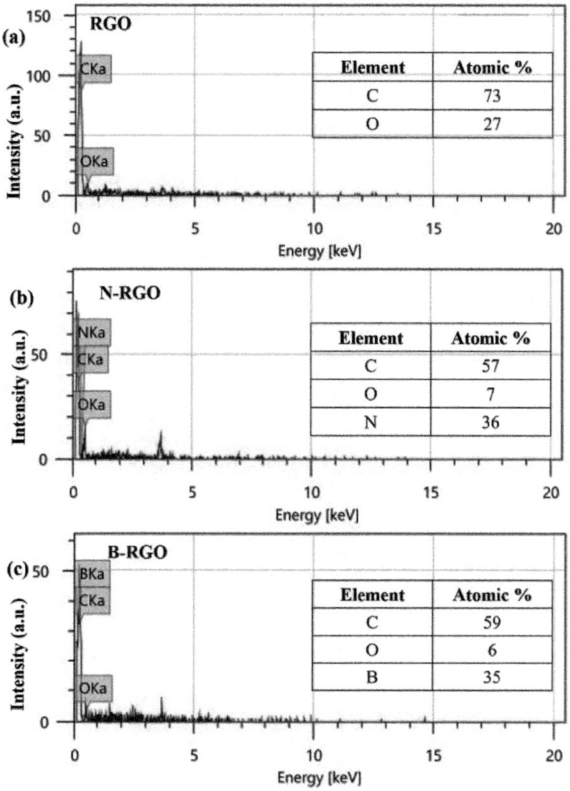

Fig. 3 Energy dispersive X-ray spectra of **a** RGO, **b** N-RGO, and **c** B-RGO

area under the CV curve of B-RGO is the largest among all. The calculated values of CS of RGO, N-RGO, and B-RGO at 10 mV/s are 212.8, 317.4, and 429.5 F/g, respectively. Figure 4b shows the GCD curves of the samples at a current density of 1 A/g, recorded in the potential window from −0.8 to 0.2 V versus Ag/AgCl. The curves have a nearly triangular symmetric shape, showing a slight linear profile during discharge. The calculated values of specific capacitance from GCD curves for RGO, N-RGO, and B-RGO are 137, 347.1, and 325.1 F/g, respectively, at 1 A/g current density. All the samples show the highest specific capacitance values at 1 A/g current density.

The CV curves of RGO, N-RGO, and B-RGO samples are also recorded at different scan rates starting from 10 to 100 mV/s. Figure 5a–c displays the CV curves of RGO, N-RGO, and B-RGO at different scan rates, and the calculated values of specific capacitance (C_S) are given in Table 1. The calculated values of C_S are highest at low scan rates and consistently decrease with an increase in the scan rate; this happens due to a relatively less time to interact with the electrode surface at high scan rates [26]. Figure 5d shows the variation of C_S values with scan rate. The capacity retentions in RGO, N-RGO, and B-RGO, while increasing the scan rate from 10 to 100 mV/s scan rates are 20.2, 52.3 and 44.6% respectively. The higher

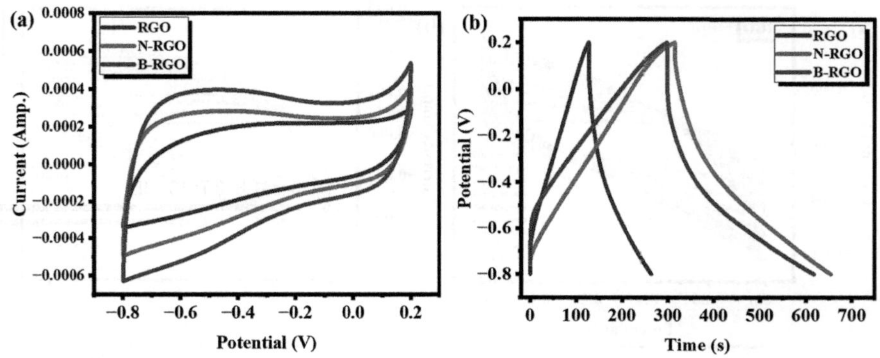

Fig. 4 **a** Cyclic voltammograms at the scan rate of 10 mV/s and **b** Galvanostatic charge–discharge curves at a current density of 1 A/g of RGO, N-RGO, and B-RGO in 1M KOH electrolyte solution

retention value in N-RGO and B-RGO shows the better accessibility of the porous electrode surface to the electrolyte ions compared to RGO. The doping of heteroatom in graphene generates defects in the sheet-like structure and further exfoliates the layers, enhancing the rate of electrochemical reactions on the electrode surface.

The specific capacitance has also been calculated from GCD curves recorded at different current densities. Figure 5e–f shows the GCD curves of N-RGO and B-RGO at 1, 2, 3, and 4 A/g; the calculated values of C_S are given in Table 2. The C_S values of N-RGO and B-RGO decrease upon increasing the current density, which is a usual phenomenon that occurs due to the low penetration rate of the electrolyte ions into the electrode's inner pore structures since the change in potential is relatively fast [27]. Both the N-RGO and B-RGO demonstrate similar values of specific capacitances (347.1 F/g and 325.1 F/g @ 1 A/g), whereas the C_S of RGO (137 F/g) is around 2.4–2.5 times less (Table 2 and Fig. 5e, f. These results further demonstrate the better results of heteroatom-doped RGO than RGO.

3.2.2 Impedance Analysis

Electrochemical impedance spectroscopy is used to analyze the samples' impedance and electrochemical charge-transfer kinetics. The Nyquist plots for the samples are shown in Fig. 6; the fitting circuit is shown in the inset image. The samples show a semi-circular plot in the high-frequency region (zoom-in image in the inset image in Fig. 6), whereas a quasi-straight line is observed in the low-frequency region. The straight line (tail region) in the low-frequency domain demonstrates an ideal capacitive behavior. The tail region is slightly bent in the form of an arc, indicating a mix of diffusive and capacitive behavior [14]. The calculated series and charge-transfer resistance (RS & RCT) values are 25 & 20 Ω for RGO, 15 & 11 Ω for N-RGO, and 14 and 8 Ω for B-RGO respectively. The R_S values for N-RGO and B-RGO are 1.7 and 1.8 times lower than RGO, while the values of R_{CT} for N-RGO and B-RGO

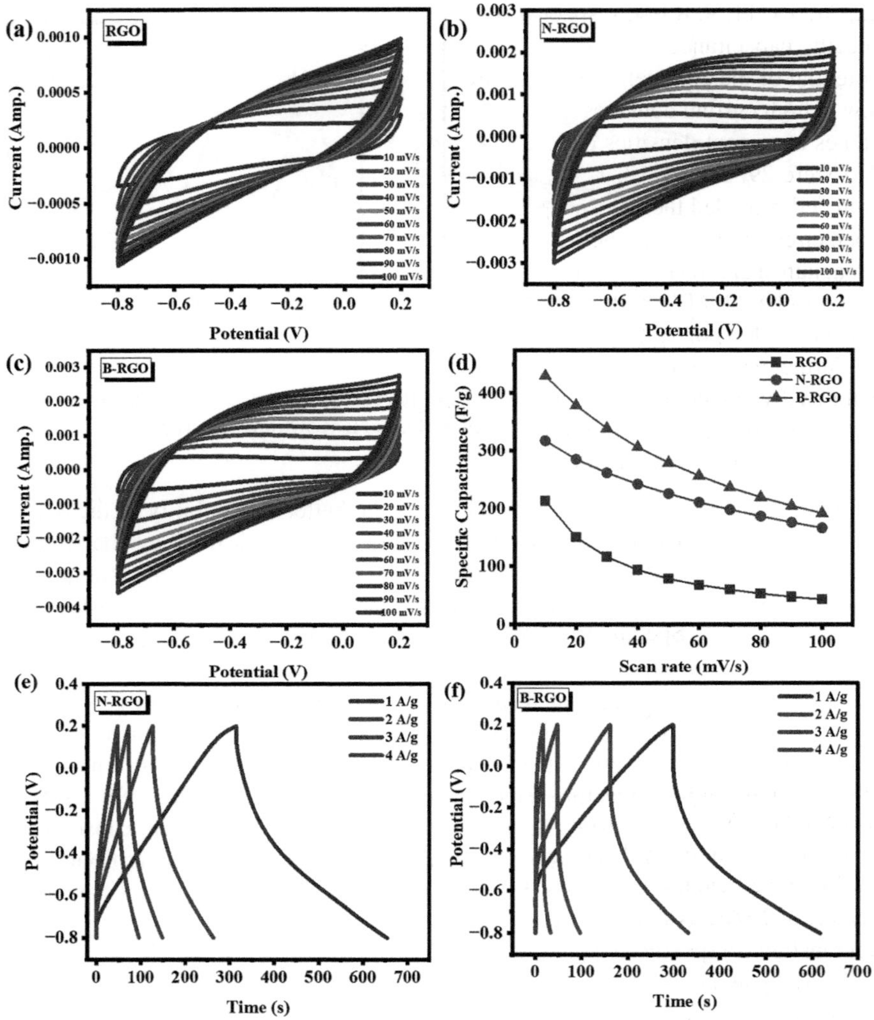

Fig. 5 Cyclic voltammograms of **a** RGO, **b** N-RGO, and **c** B-RGO at different scan rates in 1M KOH electrolyte solution in the potential range, –0.8 to 0.2 V versus Ag/AgCl. **d** the variation of specific capacitance with increasing scan rate for RGO, N-RGO, and B-RGO, **e** GCD curves of N-RGO at different current densities, **f** GCD curves of B-RGO at different current densities

are nearly 1.8 and 2.5 times lower than RGO. The R_S and R_{CT} values of N-RGO and B-RGO are significantly lower than RGO, demonstrating better charge-transfer kinetics [28].

Table 1 Calculated specific capacitance values at different current rates in the potential range, –0.8 to 0.2 V versus Ag/AgCl in 1M KOH electrolyte solution

Scan rate (mV/s)	Specific capacitance (F/g)		
	RGO	N-RGO	B-RGO
10	212.8	317.4	429.5
20	150.1	285.2	378.1
30	116.3	261.6	338.4
40	93.7	242	306.1
50	78.4	225.6	279.1
60	67.6	210.4	256.5
70	59.8	197.8	236.3
80	52.8	186.3	219.2
90	47.3	175.6	204.4
100	42.9	166.1	191.4

Table 2 Specific capacitance of RGO, N-RGO, and B-RGO at different current densities

Current density (A/g)	Specific capacitance (F/g)		
	RGO	N-RGO	B-RGO
1	137	347.1	325.1
2	67.5	146.6	178.2
3	30.6	82.2	52.5
4	15.5	53.9	23.6

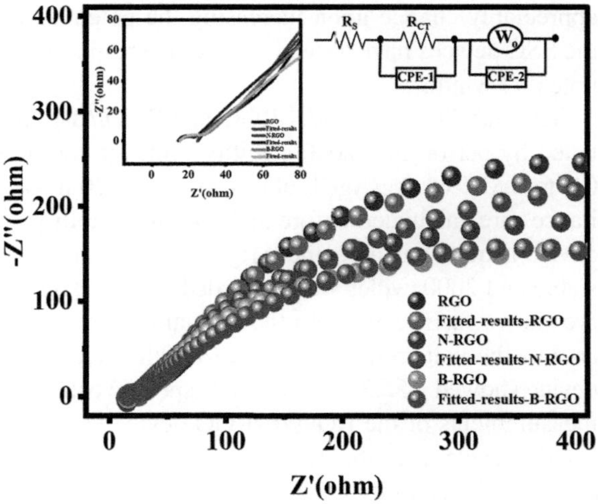

Fig. 6 Nyquist plots of RGO, N-RGO, and B-RGO with the fitted curves (The fitting circuit is shown in the inset image and the zoom-in image in the high-frequency region)

3.2.3 Device Performance

Two symmetric supercapacitor devices are assembled using RGO as anode and N-RGO and B-RGO as cathodes (RGO//N-RGO and RGO//B-RGO). A polymer soaked in the 3M KOH solution is used as the electrolyte. These symmetric supercapacitor (SSC) devices' performance is evaluated using CV and GCD in a two-electrode setup. Figure 7a, c show the CV curves of these SSCs at different scan rates in a 1.0 V wide potential window (0.0–1.0 V vs. Ag/AgCl). The area of the CV scans increases with an increase in the scan rate while the shape remains unchanged but twisted due to high currents at 0.0 and 1.0 V. The values for C_s for RGO//B-RGO and RGO//N-RGO at 10 mV/s scan rate are 172.38 and 110.2 F/g. GCD cycling is performed at different current densities to evaluate the device's reliability. Figure 7b, d show the GCD curves of RGO//B-RGO and RGO//B-RGO symmetric supercapacitor (SSC) devices. The RGO//B-RGO and RGO//N-RGO have C_S values of 133.2 and 50.5 F/g at a current density of 0.5 A/g. The specific capacitance of RGO//B-RGO and RGO//N-RGO decreases to 26 and 47%, respectively, upon increasing the current density from 0.5 to 2 A/g. The energy and power densities of the SSC devices are calculated at 0.5 A/g current density. N-RGO shows the values of energy and power densities 7.0 Wh kg^{-1} and 69.0 W kg^{-1}, whereas, B-RGO shows the values of energy and power densities 18.5 Wh kg^{-1} and 69.4 W kg^{-1}. Figure 7e, f show the CV curves of RGO//B-RGO and RGO//N-RGO supercapacitor devices, respectively, scanned from 0.0 V (vs. Ag/AgCl) to different potential potentials (0.8, 0.9, 1.0, 1.1, 1.2, 1.3, 1.4 and 1.5 V) at a scan rate of 10 mV/s. The shapes of the CV curves do not appreciably change upon increasing the potential window, this suggests that both the SSC devices have a steady capacitive response and can be used at any operating potential window.

The stability of RGO//N-RGO and RGO//B-RGO SSC devices have been evaluated by performing 2000 stability cycles in the potential window of 0.8 V (from 0.0 to 0.8 V vs. Ag/AgCl) at a scan rate of 20 mV/s. The values of specific capacitances are calculated before and after the stability cycles. The RGO//N-RGO SSC device is quite unstable under rigorous potential cycling, whereas RGO//B-RGO is stable and 2000 cycles were recorded. The first and 2000th CV cycles of the device are shown in Fig. 8a; both the voltammograms have similar shapes, however, the area of the voltammogram in the 2000th cycle decreases. The RGO//B-RGO SSC device showed a 56.3% retention of specific capacitance after 2000 cycles. The EIS measurements of the RGO//B-RGO device were carried out before and after 2000 stability cycles Fig. 8b; the series and charge-transfer resistances of the device, before and after the stability cycles are evaluated by fitting the Nyquist plots into the RC circuit. The R_S and R_{CT} values before and after cycling are 2.1 & 2.3 Ω and 134 & 330 Ω, respectively. An increase in the value of R_{CT} (134–330 Ω) upon cycling indicates the deterioration of the electrode. Corrosion of the electrode takes place with cycling which results in a reduction in the porosity and number of the active sites, which leads to an increase in the impedance [5].

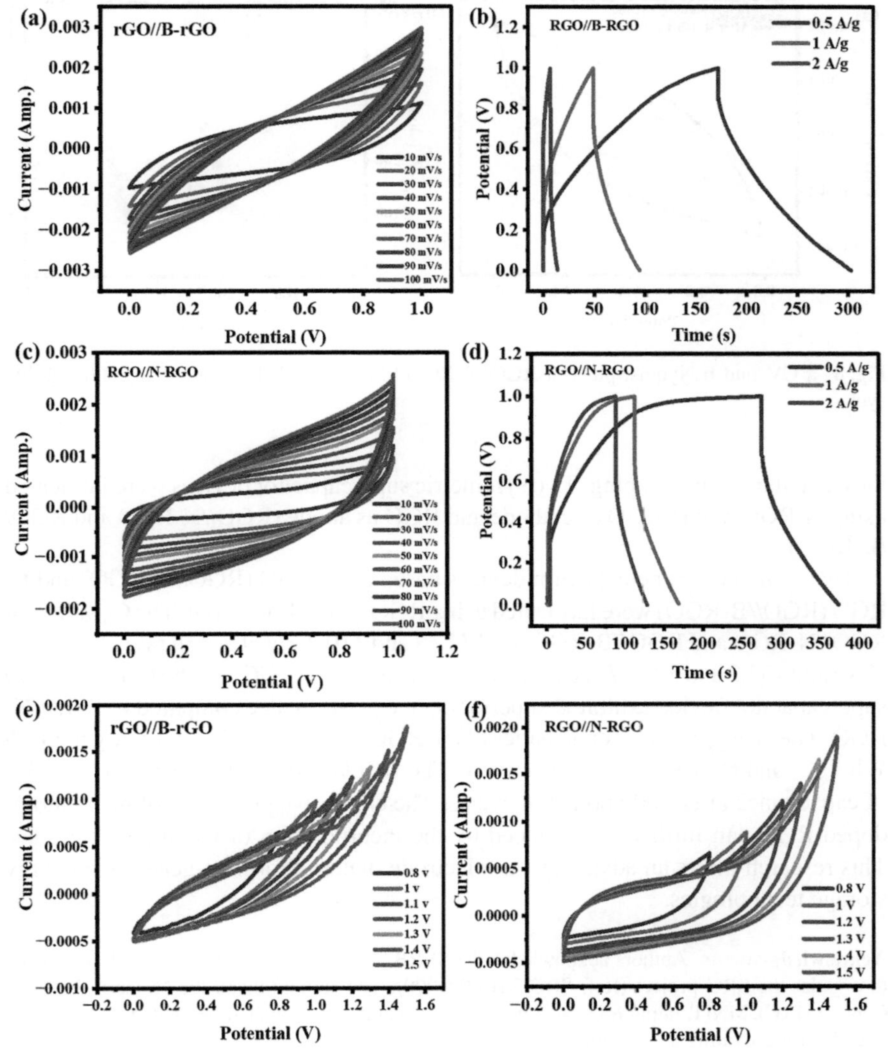

Fig. 7 **a**, **c** CV of RGO//B-RGO, RGO//N-RGO SSC devices at different scan rates, **b**, **d** GCD curves of RGO//B-RGO, RGO//N-RGO SSC devices at different current densities and **e**, **f** CV of RGO//B-RGO, RGO//N-RGO SSC devices in different potential window

4 Conclusion

Current work reports the successful doping of reduced graphene oxide with boron and nitrogen, and the evaluation of the supercapacitive behavior. The specific capacitance of RGO, N-RGO, and B-RGO at a scan rate of 10 mV/s (CV in the range of 0.0–1.0 V) are 137.2, 334.6 and 442.6 F/g, respectively, this evidence a remarkable increase in

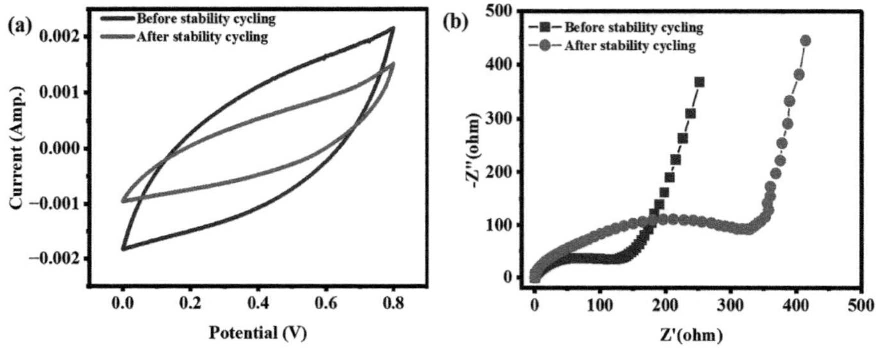

Fig. 8 **a** CV, and **b** Nyquist plots of RGO//B-RGO SSC device before and after 2000 stability cycles

the capacitance upon doping. Two symmetric supercapacitor devices were fabricated using N-RGO and B-RGO as cathode and RGO as anode (RGO//N-RGO and RGO//B-RGO).

Two symmetric supercapacitor devices based on N-RGO (RGO//N-RGO) and B-RGO (RGO//B-RGO) were fabricated using RGO as anode material. The C_S values of RGO//N-RGO and RGO//B-RGO are 172.38 and 110.2 F/g at a scan rate of 10 mV/s (CV) and 133.2 and 50.5 F/g at a current density of 0.5 A/g (GCD). The RGO//B-RGO supercapacitor device exhibited superior device performance compared to RGO//N-RGO. The energy and power densities of the RGO//B-RGO device at 0.5 A/g are 18.5 Wh kg^{-1} and 69.4 W kg^{-1}, respectively. The RGO//B-RGO device retained 56.3% of capacitance after 2000 potential cycles. The capacitive properties of heteroatom-doped RGO can further be enhanced by the incorporation of metal nanoparticles. This research offers an advantageous electrode material for next-generation energy storage technologies.

Acknowledgements Authors acknowledge the SERB, New Delhi for funding the lab infrastructure under the SRG project (Ref: SRG/2019/001090). The authors thank the Material Research Centre (MRC), MNIT, Jaipur for providing materials characterization facilities and Prof. Kanupriya Sachdev for providing lab facilities.

References

1. Guo W, Yu C, Li S, Qiu J (2021) Toward commercial-level mass-loading electrodes for supercapacitors: opportunities, challenges and perspectives. Energy Environ Sci 14:576–601. https://doi.org/10.1039/d0ee02649b
2. Şahin ME, Blaabjerg F, Sangwongwanich A (2022) A comprehensive review on supercapacitor applications and developments. Energies 15(3):674. https://doi.org/10.3390/en15030674
3. Huang S, Zhu X, Sarkar S, Zhao Y (2019) Challenges and opportunities for supercapacitors. APL Mater 7:100901. https://doi.org/10.1063/1.5116146

4. Afif A, Rahman SMH, Azad AT, Zaini J, Islan MA, Azad AK (2019) Advanced materials and technologies for hybrid supercapacitors for energy storage—a review. J Energy Storage 25:100852. https://doi.org/10.1016/j.est.2019.100852
5. Shrivastav M, Kushwaha HS, Dhiman R (2022) 2H-MoS_2 as an electrode material for oxygen reduction reaction and supercapacitor applications. Mater Today Proc. https://doi.org/10.1016/j.matpr.2022.10.202
6. Cao X, Wang X, Cui L, Jiang D, Zheng Y, Liu J (2017) Strongly coupled nickel boride/graphene hybrid as a novel electrode material for supercapacitors. Chem Eng J 327:1085–1092. https://doi.org/10.1016/j.cej.2017.07.010
7. Suresh Balaji S, Karnan M, Kamarsamam J, Sathish M (2019) Synthesis of boron-doped graphene by supercritical fluid processing and its application in symmetric supercapacitors using various electrolytes. Chem Electro Chem 6(5):1492–1499. https://doi.org/10.1002/celc.201801490
8. Johra FT, Jung WG (2015) Hydrothermally reduced graphene oxide as a supercapacitor. Appl Surf Sci 357:1911–1914. https://doi.org/10.1016/j.apsusc.2015.09.128
9. Brownson DAC, Banks CE (2016) Handbook of graphene electrochemistry. https://doi.org/10.1007/978-1-4471-6428-9
10. Lu W, Sofer Z, Simek P, Tomandl I, Pumera M (2013) Boron-doped graphene scalable and tunable p-type carrier concentration doping. J Phys Chem 117:23251–23257. https://doi.org/10.1021/jp405169j
11. Muthu RN, Sarma S, Tatiparti V (2020) Electrode and symmetric supercapacitor device performance of boron-incorporated reduced graphene oxide synthesized by electrochemical exfoliation. Energy Storage 2:e314. https://doi.org/10.1002/est2.134
12. Zhou Y, Yen CH, Fu S, Yang G, Zhu C, Du D, Wo PC, Cheng X, Yang J, Wai CM, Lin Y (2015) One-pot synthesis of B-doped three-dimensional reduced graphene oxide via supercritical fluid for oxygen reduction reaction. Green Chem 17(6):3552–3560. https://doi.org/10.1039/c5gc00617a
13. Mishra RK, Choi GJ, Sohn Y, Lee SH, Gwag JS (2020) A novel RGO/N-RGO supercapacitor architecture for a wide voltage window, high energy density and long-life via voltage holding tests. Chem Comm 56:2893–2896. https://doi.org/10.1039/d0cc00249f
14. Yadav A, Kumar R, Sahoo B (2022) Exploring supercapacitance of solvothermally synthesized N-rGO sheet: role of N-doping and the insight mechanism. Phys Chem Chem Phys 24:1059–1071. https://doi.org/10.1039/d1cp03694g
15. Li R, Lai L, Su S, Dai H, Cui Y, Zhu X (2020) One-step facile hydrothermal synthesis of heteroatom-doped porous graphene reduced comparatively by different reductants for high performance supercapacitors. Mater Today Commun 23:101118. https://doi.org/10.1016/j.mtcomm.2020.101128
16. Susanti D, Andrameda YA, Nurdiansah H, Purwaningsih H, Pradesar Y, Ardhyananta H (2019) The preparation of electrical double layer capacitor (EDLC) from boron-doped reduced-graphene oxide (B-rGO) material. IOP Conf Ser Mater Sci Eng 547:012063. https://doi.org/10.1088/1757-899X/547/1/012063
17. Mishra RK, Choi GJ, Sohn Y, Lee SH, Gwag JS (2019) Nitrogen-doped reduced graphene oxide as excellent electrode materials for high performance energy storage device applications. Mater Lett 245:192–195. https://doi.org/10.1016/j.matlet.2019.03.010
18. Kalia S, Salim A, Bala R, Singh RK, Singhal R, Dhiman R (2022) Low energy Ag ion irradiation-induced surface modification studies of Au-graphene oxide nanocomposite thin films. Thin Solid Films 772:139802. https://doi.org/10.1016/j.tsf.2023.139802
19. Jhankal D, Khan MS, Jhankal KK, Sachdev K (2022) Charge storage kinetics of MoS_2 flower decorated reduced graphene oxide for quasi solid-state symmetric supercapacitor. J Phys Chem Solids 173:111117. https://doi.org/10.1016/j.jpcs.2022.111117
20. Jiao X, Qiu Y, Zhang L, Zhang X (2017) Comparison of the characteristic properties of reduced graphene oxides synthesized from natural graphites with different graphitization degrees. RSC Adv 7:52337–52344. https://doi.org/10.1039/c7ra10809e

21. Kimbonguila A, Nkeletela LB, Abdoul-latif FM, Ndangui CB, Waiss I, Mayembo WJM, Matos L, Petit J, Scher J, Nzikou JM (2019) Effect of physical treatment on the physicochemical, rheological and functional properties of yam meal of the cultivar 'ngumvu' from dioscorea alata L. of Congo. Int J Recent Sci Res 10:30693–30695. https://doi.org/10.24327/IJRSR
22. Kumar C, Donoso B, Silva H, Padilla-Campos L, Zarate A (2021) The enhanced light trapping nature of NiOx thin films deposited by magnetron sputtering onto silicon solar cells at room temperature. Mater Lett 297:129961. https://doi.org/10.1016/j.matlet.2021.129961
23. Yadav R, Joshi P, Hara M, Yana T, Hashimoto S, Yoshimura M (2020) Intercorrelation between physical and electrochemical behavior of nitrogen-doping in graphene for symmetric supercapacitor electrode SN Appl Sci 2:1630. https://doi.org/10.1007/s42452-020-03401-x
24. Di Z, Shen H, Guo Y, Guo X, Kang B, Guo M, Wei Y, Jia J, Zhang R (2022) Heteroatom doping effect of Pt/rGO catalysts for formaldehyde abatement at ambient temperature. Chem Phys Impact 5:100103. https://doi.org/10.1016/j.chphi.2022.100103
25. Girometta C, Zeffiro A, Malagodi M, Savino E, Doria E, Neilsen E, Buttafawa A, Dondi D (2017) Pretreatment of alfalfa stems by wood decay fungus Perenniporia meridionalis improves cellulose degradation and minimizes the use of chemicals. Cellulose 24:3803–3813. https://doi.org/10.1007/s10570-017-1395-6
26. Alkhalaf S, Ranaweera CK, Kahol PK, Siam K, Adhikari H, Mishra SR, Perez F, Gupta BK, Ramasamy K, Gupta RK (2016) Electrochemical energy storage performance of electrospun $CoMn_2O_4$ nanofibers. J Alloys Compd 692:59–66. https://doi.org/10.1016/j.jallcom.2016.09.005
27. Kannappan S, Kaliyappan K, Manian RK, Pandian AS, Yang H, Lee YS, Jang J-H, Lu W (2013) Graphene based supercapacitors with improved specific capacitance and fast charging time at high current density. http://arxiv.org/abs/1311.1548
28. Magar HS, Rabeay YAH, Mulchandani A (2021) Electrochemical impedance spectroscopy (EIS): principles, construction, and biosensing applications. Sensors 21(19):6578. https://doi.org/10.3390/s21196578

A Comparison Between Wet and Dry Electrode Coating Technology for Li-Ion Battery

Anshu and Sagar Mitra

1 Introduction

The energy consumption increased by 155% between the 1990s and the 2020s, with the transportation sector seeing the largest growth. Fossil fuels are the primary conventional energy source used in transportation. According to the International Energy Agency (IEA), in 2019, world fossil fuel consumption for the transport sector was around 54%, which is very high compared to other sectors. Not only in the transport sector, but the lifestyle of the current population also requires high-energy demand, like portable electronics, vehicles, etc. The increasing energy demand seeks to develop an alternative energy source. The need for energy in this decade requires immense efforts focused on reducing the cost and enhancing the energy density of energy storage devices. The source we are searching for must be able to fulfill our demands in the transport sector, for portable electronics and others. The splendid alternative is batteries like Lithium-ion batteries (LIBs). LIBs are easy to dispose of as they contain lithium instead of cadmium; because of this and the above reasons, LIBs successfully displaced Nickel–cadmium batteries from the market of electronic devices.

As LIBs are already carved into a niche in the market of portable electronic devices, the next goal is to use them in the electrification of automobiles for widespread adoption and large-scale energy storage systems. The Department of Energy has suggested a three to four times reduction in the current costs system, i.e., \$152 kWh^{-1} needed to meet the requirements for widespread acceptance. Different technologies are trying to get involved in this global cost-reduction effort by developing various methods for the fabrication technology of electrodes. As of early 2021, the average cost of a LIBs pack for electric vehicles was around \$137/kWh, according

Anshu (✉) · S. Mitra
Department of Energy Science and Engineering, Indian Institute of Technology Bombay, Mumbai, Maharashtra 400076, India
e-mail: anshu52iitb@gmail.com

S. T. Aruna et al. (eds.), *Recent Advances in Electrochemical Science and Technology*, Springer Proceedings in Materials 47, https://doi.org/10.1007/978-981-97-3958-5_6

to Bloomberg NEF. This represents a significant decline from the average price of \$1,100/kWh in 2010. It is expected to decline further in the coming years as the demand for electric vehicles increases and technology continues to improve. This rapid decline in the cost of LIBs is due to technological advancements in electrode fabrication, the economics of scale, and increased production.

Compared with the other components, electrodes are a critical component of LIBs, and their performance and manufacturing significantly impact the overall performance and costs of the battery. The most used electrode in LIBs was manufactured using a conventional wet electrode coating technology. It involves mixing the active material, binder, and additive in a solvent. The disadvantage of using a solvent is its toxicity and demand for higher temperatures for evaporation. So, the alternative method discovered by researchers to manufacture electrodes is the dry battery electrode (DBE) technology. DBE offers significantly high loading, avoids using solvents, and is environment friendly. This review provides recent progress in electrode manufacturing and compares the performance, manufacturing cost, and durability of electrodes made by dry electrode coating processes and wet slurry coating processes.

2 Electrode Manufacturing

An electrode is where the electrochemical reactions throughout the charging and discharging process occur, which is a crucial part of a battery. As a vital part of a battery, an electrode is essential to the storage and discharge of the battery. The electrodes in a lithium battery pack comprise the largest percentage of the pack's weight, accounting for around 45–50% [1, 2]. It has been observed that the contribution of electrodes in cost and by weight in the battery is comparatively higher than other components.

As an essential part of the battery, electrode manufacturing is also a subject of discussion due to its overall participation in battery performance and cost. The commonly used technology in electrode manufacturing is wet battery electrode coating technology.

2.1 Wet Electrode Coating Technology

The conventional wet electrode manufacturing process consists of mixing, coating, drying, calendaring, post-drying, and cell assembly steps, as shown in Fig. 1 [2, 3]. The wet process follows the essential step of a slurry formation consisting of active materials, binders, conductive additives, and solvents. The classical slurry in wet coating technology contains a binder, commonly polyvinylidene fluoride (PVDF), carbon as conductive additives, and active material like Lithium metal oxide, which is dissolved in solvent N-Methyl-2-pyrrolidone (NMP) [4, 5]. This uniformly mixed slurry is now coated onto a substrate, the current collector, most commonly used copper and aluminum, and then dried and pressed to form an electrode [6–8].

NMP solvent used in the wet coating process must be recovered or removed by evaporation because NMP has a high cost and pollutes the environment. Solvent evaporation and rate of evaporation have a large impact on the redistribution of each component of slurry and also on electrode microstructure [9, 10] (Fig. 2).

NMP evaporation is performed at temperatures over 120 °C and a long-range time, requiring significant energy in the form of electricity [11]. The electricity used to provide energy for solvent evaporation further increases the cost of batteries and the consumption of fossil fuels. The recovery system makes the process environmentally

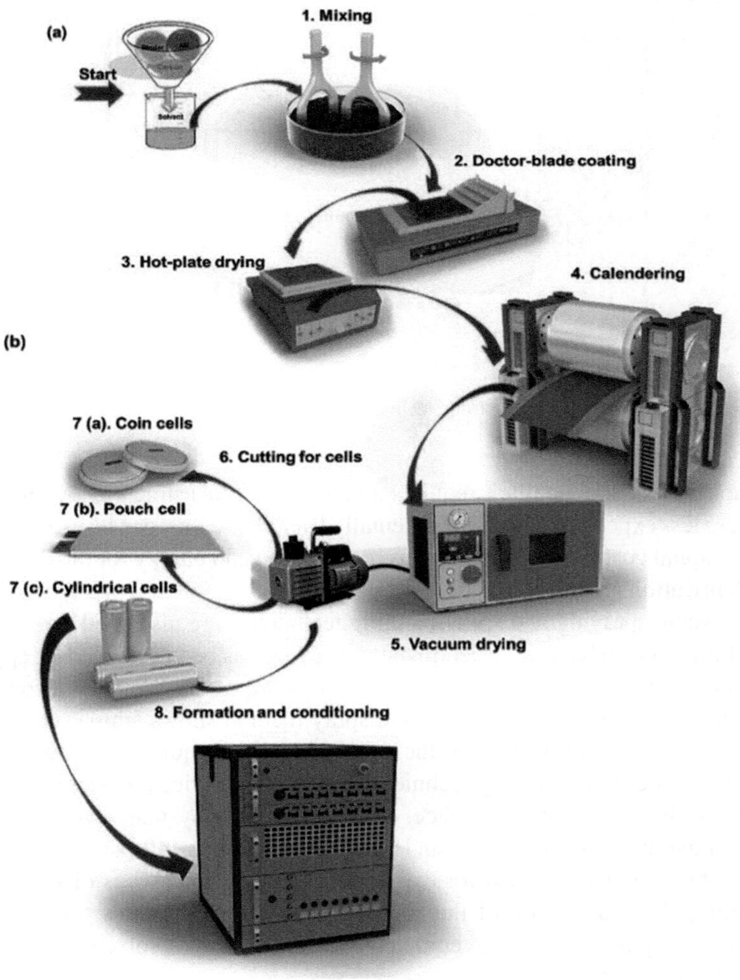

Fig. 1 Steps of electrode and battery manufacturing process. Adapted with permission [1]. Copyright 2021, Wiley

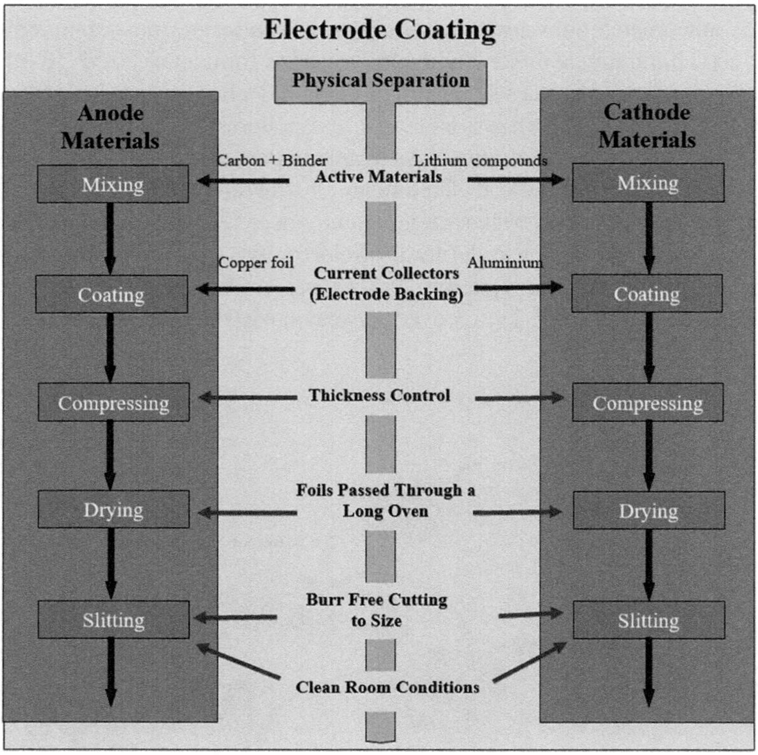

Fig. 2 Schematic of wet-coated electrode technology

sustainable, but economically it requires a large capital investment. Aqueous-based slurries are less expensive and environmentally friendly, so they can be used to reduce the large capital cost. But still, electrodes require time and energy for the drying step during fabrication [12].

Electrostatic spraying and spray-painting techniques are also used for the fabrication of lithium-ion batteries. Electrostatic spraying has been achieved by applying DC voltage between an electrically conductive metal and a capillary nozzle [13]. The electrode formed by the electrostatic spray method shows similar characteristics as conventional electrodes and the disadvantage of requiring time and energy for drying. In the spray-painting technique, LIBs were fabricated using an airbrush to paint the desired substrate surface. This is achieved by maintaining substrate surface temperature at 90–120 °C and airbrush operated at 40 psi of compressed air [14]. This technique is used for the fabrication of various types of surfaces but still requires the evaporation of the solvent. An advanced process for fabricating electrodes is required, which can completely avoid the use of solvent, which is dry coating electrode technology.

One advantage of wet electrode coating technology is that it allows for using a wider range of active materials and binders and the ability to tune the slurry properties to achieve optimal electrode performance [15]. The wet coating also enables the production of thicker electrodes, resulting in higher energy–density batteries. However, using solvents in the wet coating can result in environmental and safety concerns, and the drying and pressing steps can increase the processing time and cost [16–18].

2.2 *Dry Electrode Coating Technology*

The ideal manufacturing technology that meets today's needs and everyone's expectations toward cost reduction, environmentally friendly, and energy savings with eliminating inadequacies of the conventional wet process is dry electrode coating technology. This process utilizes dry particles and is a perfect alternative solution for the replacement of the wet slurry casting process, as it removes the cost of solvent as well as the cost of their removal and recovery [11, 19]. Dry electrode coating technology consists of a dry mixture of active material, conductive additives, and binder. This dry mixture is properly mixed and then directly coated onto a substrate [20, 21]. In dry electrode coating technology, the coating of the dry mixture onto a current collector has been achieved through various methods such as pulsed laser, sputtering deposition dry pressing, and electrostatic spraying [8, 22–25].

The pulsed laser process uses a high-energy laser beam to selectively heat and evaporate thin layers of electrode materials before depositing them onto a substrate to form a thin, uniform coating. The laser beam is pulsed at a high frequency, often in the hundreds of kHz to MHz range, to produce a sequence of fast heating and cooling cycles, resulting in a plasma of vaporized material [22]. The plasma is then directed onto the substrate, where it condenses and solidifies to create an electrode material, and the process is repeated numerous times until the required thickness and structure of the electrode are attained [22, 26]. For annealing, the deposited film must be exposed to very high temperatures (650–800 °C), comparable to solvent evaporation and recovery of the wet slurry process [6] (Fig. 3).

In the sputtering deposition, a high-energy plasma is created in a vacuum chamber using an inert gas such as argon. Annealing done by sputtering can lower the temperature to 350 °C [24, 25]. These methods can be used for dry electrode coating technology but have drawbacks such as a slow deposition rate and high temperature needs for annealing [27]. A solution for depositing dry electrode particles on the collector has been realized using dry pressing and electrostatic spraying methods [13, 28]. The dry pressing technique is solely applicable to dry electrode coating technology. While other methods can be used for wet and dry battery electrode technology, the dry pressing method includes using a hydraulic press to compress dry electrode material into the required shape and density.

The electrode that results is then trimmed to the proper size and shape. In this way, it produces electrodes with high density and consistent porosity, which can

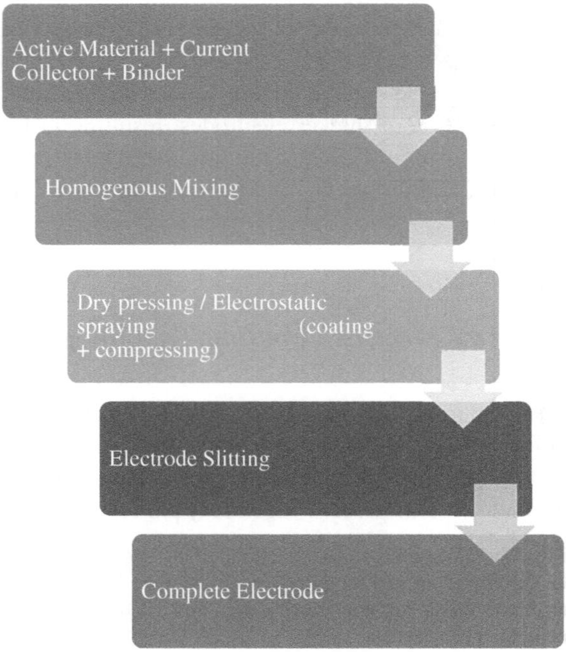

Fig. 3 Steps of dry electrode coating technology

increase electrode performance and cycle life [8]. Furthermore, the technique is straightforward and may be quickly scaled up for large-scale production. However, the dry pressing procedure has limitations, such as numerous compression steps that may be required to get the correct density and form [28]. Another method is dry painted or electrostatic spraying can be used for the deposition of electrode particles. A charged spraying gun filled with fluidized dry particles will be drawn to the current collector and deposited. The density and thickness of the electrode are maintained by using the hot roller, while in wet-coated electrodes, doctor blades are used for this [29]. Using the hot roller, activation of the binding material can be achieved quickly, which replaces the use of the oven required to evaporate the solvent in the wet coating process [2, 30]. Electrostatic spraying manifests good flexibility, and also it has the capability to coat the electrode with large surface areas [11].

The dry electrode coating technology eliminates the need for solvents and drying steps, resulting in a more environmentally friendly and cost-effective process. The dry coating also allows for better control over the thickness and uniformity of the electrode, resulting in improved battery performance.

3 Electrochemical Characterization

The electrochemical and mechanical bonding behavior both control the properties of the electrode in the battery. SEM images of dry-painted and slurry casted electrodes were observed forming both electrodes by mixing the same composition of active (90% by weight), conductive (5% by weight), and binding material (5% by weight) except solvent. Solvent only present in case of slurry cast [6].

Only a few morphological changes are observed before and after the coating of the dry coating process, as shown in Fig. 4a, b SEM image ensures the particle remains intact after coating.

Figure 4c, d shows that after calendaring, there is an even and dense distribution of electrode material on the surface in the case of dry coating electrode technology; on the other hand, the agglomeration of particles observed in the case of a wet process which decreases the contact surface area and reaction as well [11]. The electrode formed by the dry-painted process shows stronger bonding compared with the wet electrode process.

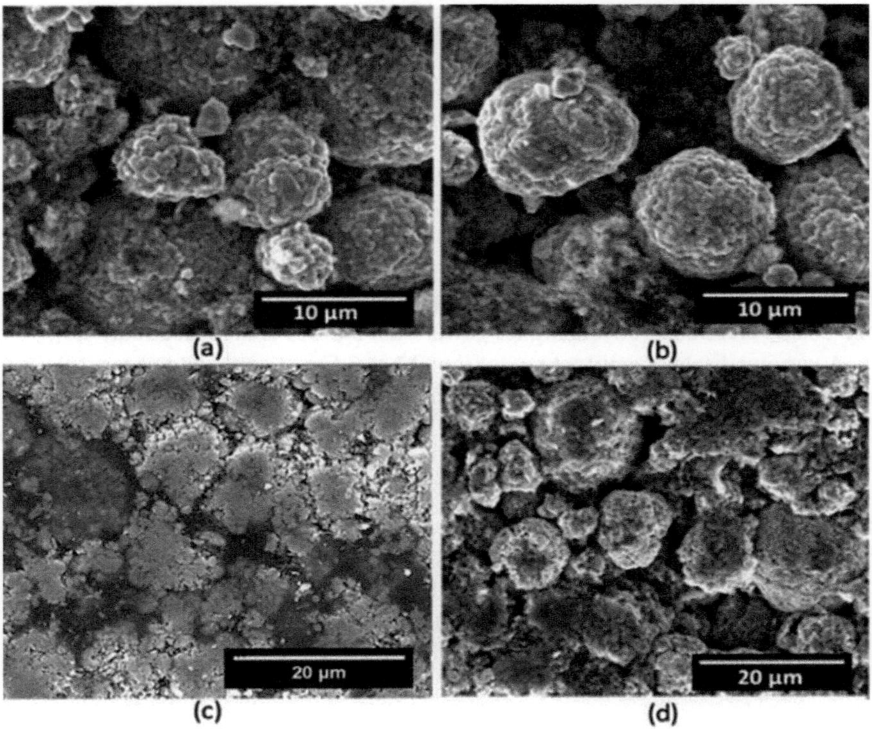

Fig. 4 SEM images of **a** Powder mixture before dry-powder coating. **b** Powder mixture after dry-powder coating. **c** Dry-powder and **d** Wet slurry-coated electrode after calendaring. Adapted with permission [11]. Copyright 2017, Elsevier

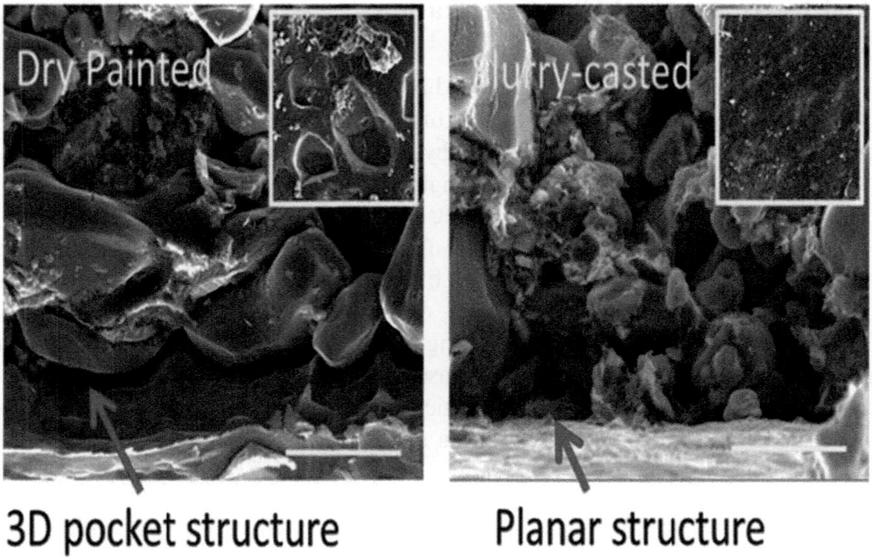

Fig. 5 SEM image comparing wet and dry electrode coating electrode. Adapted with permission [6]. Copyright 2016, Scientific Reports

The bonding at the interface of the electrode-current collector might be weaker because of the 2D planar structure in the conventional method.

In Fig. 5, the pocket structure observed on current collectors provides additional contact area and enhances adhesion strength for the dry process, while the wet process lacks this. These pocket structures formed on current collectors due to mechanical pressing used in the dry process. Once the discharge current or C-rate increases, the dry-coated electrode delivers higher power than the wet-coated electrode [20].

A varying discharge current range from 0.1 to 3 C was applied to observe the rate performance of dry-painted and conventional slurry casted electrodes in Fig. 6. It is observed that at a lower discharge rate, both the electrode has a similar capacity of 121 mAh g^{-1}, but on increasing current rate, dry-painted electrode retains higher capacity of 87 mAh g^{-1}. The dry-painted electrode has 70% capacity retention after 50 cycles, while the electrode formed by the conventional method only possesses 58% capacity retention.

Overall, the electrode formed using a dry-painted electrode has a larger capacity and higher cycling stability than the standard slurry-cast electrodes [20, 29].

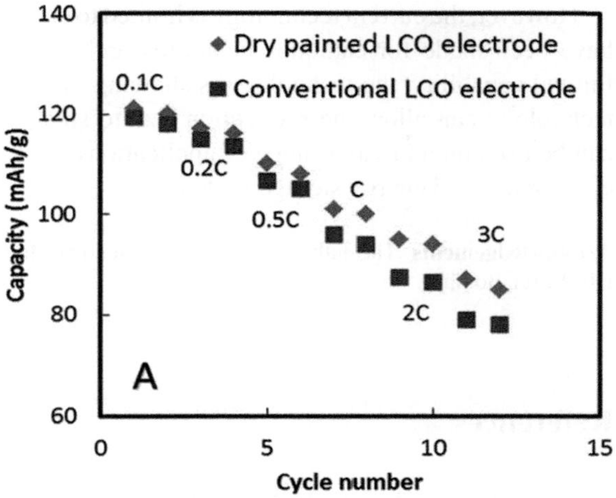

Fig. 6 C-rate performance of dry-painted electrode vs slurry casted electrode. Adapted with permission [6]. Copyright 2016, Scientific Reports

4 Conclusion

In this review, the dry battery electrode coating technology has been considered and the effect of different methods on the coating and properties of resultant electrodes were discussed. A critical review of different fabrication methods and coating processes is included with a view to understand the manufacturing dynamics. It has been observed that the dry electrode coating technology produces electrodes with high density and uniform porosity, which can improve the electrode's performance and cycle life. By eliminating the need for solvent-based coatings, reducing waste, and enabling faster production, this technology can help manufacturers produce high-quality batteries at lower costs. The process is relatively simple and can be easily scaled up for large-scale production.

As the demand for high-performance and sustainable batteries continues to grow, further advancements in dry battery electrode coating technology can be expected to improve the performance, efficiency, and safety of lithium-ion batteries.

5 Future Prospects

The dry battery electrode coating technology has shown great promise for the manufacturing of lithium-ion battery electrodes. The dry battery electrode coating technology may also lead to the creation of new materials for use in lithium. The technology can enable the production of high-quality, uniform electrodes with a wide range of materials, opening up new possibilities for battery performance and applications. It can reduce the manufacturing unit's size and the processing and drying time of cell manufacturing.

However, the current technology is limited to anode electrodes, and many attempts have been made for cathodes in the research and development sectors. For such limited capability, the technology is showing great promise in the near future. The technology can allow the production size to smaller, more powerful batteries that can be used on a broader range of applications, such as electric vehicles, consumer electronics, and energy storage systems.

Acknowledgements The authors thank research scholar Mr. Manoj Gautam for his kind support, DESE, IIT Bombay.

References

1. Al-Shroofy M, Zhang Q, Xu J, Chen T, Kaur AP, Cheng YT (2017) Solvent-free dry powder coating process for low-cost manufacturing of $LiNi_{1/3}Mn_{1/3}Co_{1/3}O_2$ cathodes in lithium-ion batteries. J Power Sources 352:187–193. https://doi.org/10.1016/j.jpowsour.2017.03.131
2. Chen CH, Kelder EM, Jak MJG, Schoonman J (1996) Electrostatic spray deposition of thin layers of cathode materials for lithium battery. Solid State Ionics 86–88:1301–1306. https://doi.org/10.1016/0167-2738(96)00305-0
3. Chiu KF (2007) Lithium cobalt oxide thin films deposited at low temperature by ionized magnetron sputtering. Thin Solid Films 515:4614–4618. https://doi.org/10.1016/j.tsf.2006.11.073
4. Dominko R, Gaberscek M, Drofenik J, Bele M, Pejovnika S (2001) A novel coating technology for preparation of cathodes in Li-Ion batteries. Electrochem Solid-State Lett 4:A187. https://doi.org/10.1149/1.1407995
5. Duan Z, Wu Y, Lin J, Wang L, Peng DL (2022) Thin-film lithium cobalt oxide for lithium-ion batteries. Energies 15(23):8980. https://doi.org/10.3390/en15238980
6. Duong H, Shin J, Yudi Y (2018) Dry electrode coating. https://api.semanticscholar.org/CorpusID:201928996
7. Eskra MD, Ralston PK (2015) Scale up of dry process for lithium ion electrode manufacturing scale up for lithium ion electrode manufacturing. https://ndiastorage.blob.core.usgovcloudapi.net/ndia/2015/power/17834eskra.pdf
8. Julien C, Camacho-Lopez MA, Escobar-Alarcon L, Haro-Poniatowski E (2001) Fabrication of $LiCoO_2$ thin-film cathodes for rechargeable lithium microbatteries. Mater Chem Phys 68:210–216. https://doi.org/10.1016/S0254-0584(00)00372-2
9. Kirsch DJ, Lacey SD, Kuang Y, Pastel G, Xie H, Connell JW, Lin Y, Hu L (2019) Scalable dry processing of binder-free lithium-ion battery electrodes enabled by holey graphene. ACS Appl Energy Mater 2:2990–2997. https://doi.org/10.1021/acsaem.9b00066
10. Lin Y, Moitoso B, Martinez-Martinez C, Walsh ED, Lacey SD, Kim J-W, Dai L, Hu L, Connell JW (2017) Ultrahigh capacity lithium-oxygen batteries enabled by dry-pressed holey graphene air cathodes. Nano Lett 17(5):252–3260. https://doi.org/10.1021/acs.nanolett.7b00872
11. Liu D, Chen L-C, Liu T-J, Fan T, Tsou E-Y, Tiu C (2014) An effective mixing for lithium ion battery slurries. Adv Chem Eng Sci 04:515–528. https://doi.org/10.4236/aces.2014.44053
12. Liu H, Cheng X, Chong Y, Yuan H, Huang JQ, Zhang Q (2021) Advanced electrode processing of lithium ion batteries: a review of powder technology in battery fabrication. Particuology 57:56–71. https://doi.org/10.1016/j.partic.2020.12.003
13. Ludwig B, Zheng Z, Shou W, Wang Y, Pan H (2016) Solvent-free manufacturing of electrodes for lithium-ion batteries. Sci Rep 6. https://doi.org/10.1038/srep23150
14. Madian M, Eychmüller A, Giebeler L (2018) Current advances in TiO_2-based nanostructure electrodes for high performance lithium ion batteries. Batteries 4(1):7. https://doi.org/10.3390/batteries4010007

15. Matsuda Y, Kuwata N, Kawamura J (2018) Thin-film lithium batteries with 0.3–30 μm thick $LiCoO_2$ films fabricated by high-rate pulsed laser deposition. Solid State Ion 320:38–44. https://doi.org/10.1016/j.ssi.2018.02.024
16. Mishra A, Mehta A, Basu S, Malode SJ, Shetti NP, Shukla SS, Nadagouda MN, Aminabhavi TM (2018) Electrode materials for lithium-ion batteries. Mater Sci Energy Technol 1:182–187. https://doi.org/10.1016/j.mset.2018.08.001
17. Mukherjee S, Mujib S Bin, Soares D, Singh G (2019) Electrode materials for high-performance sodium-ion batteries. Materials 12(12):1952. https://doi.org/10.3390/ma12121952
18. Mun J, Park JH, Choi W, Benayad A, Park JH, Lee JM, Doo SG, Oh SM (2014) New dry carbon nanotube coating of over-lithiated layered oxide cathode for lithium ion batteries. J Mater Chem A Mater 2:19670–19677. https://doi.org/10.1039/c4ta04818k
19. Reynolds CD, Slater PR, Hare SD, Simmons MJH, Kendrick E (2021) A review of metrology in lithium-ion electrode coating processes. Mater Des 209:109971. https://doi.org/10.1016/j.matdes.2021.109971
20. Ruffo R, Wessells C, Huggins RA, Cui Y (2009) Electrochemical behavior of $LiCoO_2$ as aqueous lithium-ion battery electrodes. Electrochem Commun 11:247–249. https://doi.org/10.1016/j.elecom.2008.11.015
21. Shen X, Zhang XQ, Ding F, Huang JQ, Xu R, Chen X, Yan C, Su FY, Chen CM, Liu X, Zhang Q (2021) Advanced electrode materials in lithium batteries: retrospect and prospect. Energy Mater Adv 4:1–15. https://doi.org/10.34133/2021/1205324
22. Singh N, Galande C, Miranda A, Mathkar A, Gao W, Reddy ALM, Vlad A, Ajayan PM (2012) Paintable battery. Sci Rep 2. https://doi.org/10.1038/srep00481
23. Stein M, Mistry A, Mukherjee PP (2017) Mechanistic understanding of the role of evaporation in electrode processing. J Electrochem Soc 164:A1616–A1627. https://doi.org/10.1149/2.1271707jes
24. Tintignac S, Baddour-Hadjean R, Pereira-Ramos JP, Salot R (2014) Electrochemical properties of high rate bias sputtered $LiCoO_2$ thin films in liquid electrolyte. J Power Sources 245:76–82. https://doi.org/10.1016/j.jpowsour.2013.06.047
25. Wixom M (2015) Dry process electrode fabrication 1. Al-Shroofy M, Zhang Q, Xu J, Chen T, Kaur AP, Cheng YT (2017) Solvent-free dry powder coating process for low-cost manufacturing of $LiNi_{1/3}Mn_{1/3}Co_{1/3}O_2$ cathodes in lithium-ion batteries. J Power Sources 352:187–193. https://doi.org/10.1016/j.jpowsour.2017.03.131
26. Wu X, Pan K, Jia M, Ren Y, He H, Zhang L, Zhang S (2019) Electrolyte for lithium protection: from liquid to solid. Green Energy Environ 4:360–374. https://doi.org/10.1016/j.gee.2019.05.003
27. Yan B, Liu J, Song B, Xiao P, Lu L (2013) Li-rich thin film cathode prepared by pulsed laser deposition. Sci Rep 3. https://doi.org/10.1038/srep03332
28. Zhang W, He X, Pu W, Li J, Wan C (2011) Effect of slurry preparation and dispersion on electrochemical performances of $LiFePO_4$ composite electrode. Ionics 17:473–477. https://doi.org/10.1007/s11581-011-0560-4
29. Zhang YS, Bailey JJ, Sun Y, Boyce AM, Dawson W, Reynolds CD, Zhang Z, Lu X, Grant P, Kendrick E, Shearing PR, Brett DJL (2022) Applications of advanced metrology for understanding the effects of drying temperature in the lithium-ion battery electrode manufacturing process. J Mater Chem A Mater. https://doi.org/10.1039/d2ta00861k
30. Zhang YS, Courtier NE, Zhang Z, Liu K, Bailey JJ, Boyce AM, Richardson G, Shearing PR, Kendrick E, Brett DJL (2022) A review of lithium-ion battery electrode drying: mechanisms and metrology. Adv Energy Mater 12:2102233. https://doi.org/10.1002/aenm.202102233

Flexible IDE-Based ZnO/MWCNT Composite Sensor for Real-Time Nitric Oxide Monitoring

Pratima Bhat and S. S. Omprakash

1 Introduction

Environment protection is a global issue. Sustainable development requires that a country develops monitoring strategies for ensuring the acceptable background pollution levels. In particular, toxic gases such as nitric oxide (NO_x), carbon monoxide, methane, sulphuric oxide, and volatile organic compounds (VOCs) like benzene, toluene, etc., must be monitored on real time, as their inhalation leads to severe heart or lung diseases [1, 2]. Gas sensors play a major role in environmental aspects due to concerns about pollution, health, and safety. Research on developing new materials with improved diffusion, high sensitivity under ambient conditions, and low power consumption has been shaped by the increasing need for sensitive, quick, and low-cost gas sensors in industrial environment, medical and clinical diagnostics and automotive applications [3, 4].

Sensors based on semiconducting metal oxides like ZnO have tunable responsivity to target analyte, fast response and recovery time, convenient in operation, when synthesized in hollow and porous forms like nanorods, nanoflowers, nanosheet, etc., when compared with SnO_2, TiO_2, WO_3, NiO, CuO, etc. [5–8].

The technological challenge has led to the fabrication of composite gas sensors obtained by the addition of some functional materials like CNT (Carbon Nanotube). Their unique geometry, morphology like fullerene structure, electronic, mechanical, and thermal properties make them ideal for advanced gas sensing applications [9–11]. Semiconducting MWCNTs are extremely sensitive to charge transfer in the presence of oxidizing analyte [12] which is due to its tube size and the chiral structure. The high van der Waals force, due to the presence of π-π bond, is considered to be the

P. Bhat (✉)
Department of Electronics, Jyoti Nivas College Autonomous, Bangalore, Karnataka, India
e-mail: prathimabhat@jyotinivas.org

S. S. Omprakash
Department of Physics, Prayoga Institute of Educational Research, Bangalore, Karnataka, India

S. T. Aruna et al. (eds.), *Recent Advances in Electrochemical Science and Technology*, Springer Proceedings in Materials 47, https://doi.org/10.1007/978-981-97-3958-5_7

main challenge since the structure is ideal for the hydrophobic interactions with the analyte which leads to long recovery times. The dispersion of MWCNT is formidable due to its poor solubility in solvents like water, DMSO (dimethyl sulfoxide), and chloroform.

Sabri et al. [13] have studied the electronic characterization of MWCNT coated glass fiber via electrospray deposition when dispersed using various solvents like NMP (N-methyl pyrrolidone), DMF (Dimethylformide), and THF (tetrahydrofuran). Marcel et al. [14] have studied the dispersion of MWCNT in poly (vinylidene fluoride) (PVDF) in different wt% solutions and explored α and β piezoelectric phase of PVDF to produce polymer nanocomposites. To overcome agglomeration, surface functionalization, or addition of surfactants like sodium dodecyl sulfate and Triton –X 100 can be added [15, 16].

Interdigitated electrodes are fabricated to analyze the electrical signals from the sensor. They consist of a few parallel micro-leg electrodes closely spaced with a contact pad. IDE can be developed as a bottom or top contact for a sensor. Bottom contact is beneficial as it forms maximum exposed area for the analyte to interact [17, 18]. IDE can be developed using methods like photolithography, thermal evaporation, stamping, screen printing, etc. [19].

Nitric oxide is the by-product released from many chemical industries and vehicle emissions which contributes to the pollution level. Nitric oxide (NO) is one of the toxic gases that adversely impacts the environment, causing acid rain, global warming by destruction of ozone, water quality degradation, and visual impairment. Inhaling high levels of NO can damage the human respiratory tract and make a person more susceptible to respiratory infections and asthma. Higher concentration of NO and prolongs exposure to NO can lead to lung cancer. In clinical diagnostics, nitric oxide (NO) is considered as an important biomarker for the early diagnosis of asthma or respiratory diseases.

In the present study, ZnO/f-MWCNT composite material is synthesized using wet chemical method and studied for nitric oxide sensing to analyze the parameters like sensitivity, response, and recovery time. For integration of the gas sensor with the gas chamber, designing of IDE-based structure is also discussed followed by the instrumentation system.

2 Materials and Methods

2.1 Preparation of Sensing Layer and Characterization

ZnO/f-MWCNT composite was synthesized using a wet chemical process. Zinc acetate dihydrate was used as the basic precursor to form a solution in alcohol medium. Potassium hydroxide was added dropwise to this precursor solution in 20:1 ratio to maintain the pH level, during the growth of ZnO nanostructure. Functionalized MWCNT (length 15 μm and 50 nm in diameter) was procured from

Avansa Technologies, India. 22 mg of procured f-MWCNT is added to 20 ml of DMF (Dimethyl formide) solution and ultrasonicated at 60 °C for 20 min for proper dispersion. DMF, an organic polar solvent, is hydrophilic in nature and is preferred for the dispersion of MWCNT. This solution is further added dropwise to as prepared zinc oxide sol during its growth. The mixture solution is stirred in a reflux system at 160 °C for an hour followed by centrifugation to obtain composite.

The composite was further annealed at 400 °C for one hour. The annealing process is effective in accelerating the adsorption and desorption of gas molecules on the surface.

Thin film coating was obtained by spin coating (composite) the sol at 3000 rpm for 20 s on the substrate. Flexible Teflon substrate was chosen and the sensing layer was coated on the IDE electrode forming a bottom contact for the sensor. Surface thickness and crystallographic growth structures of the composite material were studied using atomic force microscope (AFM), X-ray diffractometer (XRD), and scanning electron microscope (SEM) respectively. The crystallite size (D) of the composite was calculated using Debye–Scherer's equation [20].

2.2 Design of Mask and Development of Interdigitated Electrode Structure

An interdigitated electrode (IDE) structure was developed (Fig. 1b) to integrate sensor within the chamber and to pick the electrical variations of the sensing layer on exposure to the target gas. A stainless steel mask is designed and developed using CAD software with dimensions as shown in Fig. 1a.

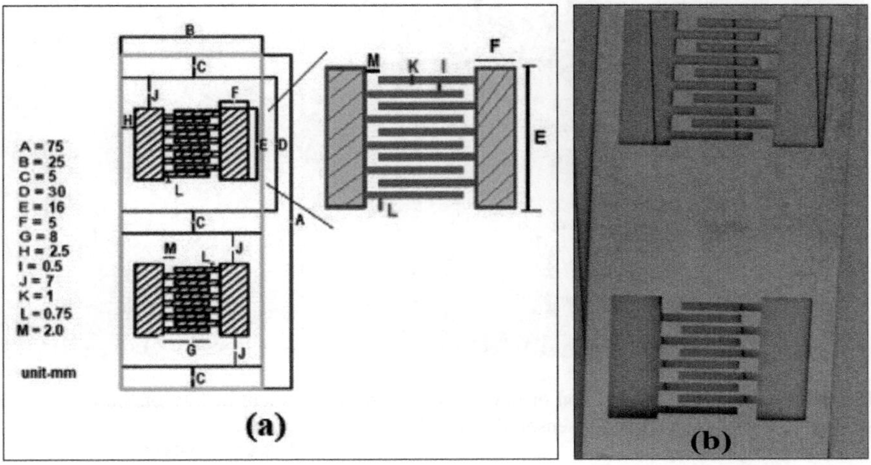

Fig. 1 **a** CAD structure of IDE, **b** schematic of the developed mask

Interdigitated electrodes with narrow width that maximizes the contact area of the signal being measured and the sensing film (width between the electrodes are kept at 0.5 mm). With the help of a mask, an aluminum-based IDE was developed on the substrate using thermal evaporation, upon which the sensing layer is coated using spin coating technique. Schematic representation of the sensor development process is shown in Fig. 2a. Aluminum metal developed as a bottom electrode on the surface of different substrates like glass/flexible is shown in Fig. 2b. From this electrode, contact wires are taken out as shown in Fig. 2c with the help of silver paste over which the sensing layer (composite sol) is coated as a thin film.

Due to the notable advantages of the flexible substrate like lightweight, bending flexibility, proximity use, low power intake, efforts are made to develop flexible nanosensors as breath analyzers at a lower cost in health care applications. Sensing layer was coated as a thin film over this electrode using spin coating technique as shown in Fig. 2d.

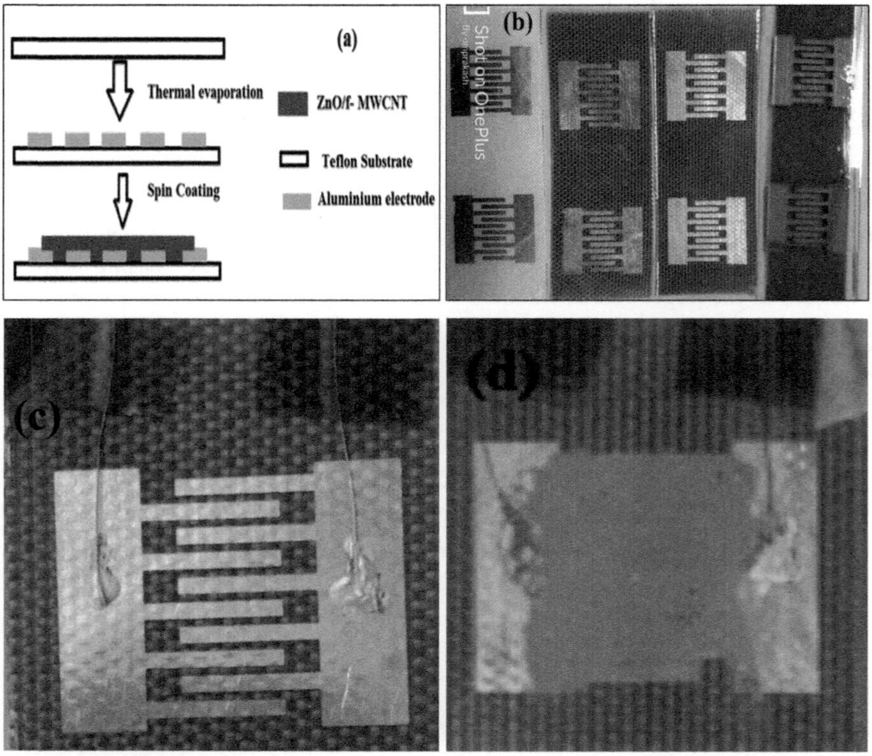

Fig. 2 **a** Schematic representation of devised sensor, **b** development of IDE on various flexible substrates, **c** wire connection for sensor, **d** ZnO/f-MWCNT film on IDE

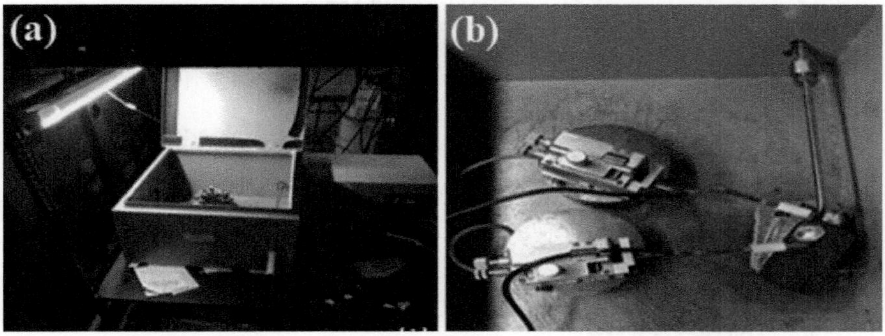

Fig. 3 **a** Gas sensing chamber, **b** inner view—sensor with electrical assembly

2.3 Gas Sensing Setup

The gas sensing studies were performed in a closed chamber connected with an electrometer, Keithley 2450, shown in Fig. 3. It is an assembled system that has a substrate holder with a heater attached to it that determines the working temperature of the sensor. The source meter was connected to the contact pad of the IDE film to measure the electrical variations of the film in terms of current upon exposure to the target gas.

NO gas was the target gas and synthetic air was taken as carrier gas. Both the gases were connected to the gas chamber through a mass flow controller. 500 sccm (standard cubic centimeter per minute) of NO and synthetic gas were allowed to pass through the mass flow controller in order to obtain 2.5 ppm of nitric oxide inside the chamber.

3 Results and Discussions

3.1 Characterization

XRD pattern of ZnO, MWCNT, and ZnO/f-MWCNT composite is shown in Fig. 4a. The highest intensity peak for the composite sample was observed at a Bragg's angle of $2\theta = 36.33°$, corresponding to (101) hkl plane. All the peaks of the composite match with that of ZnO having JCPDS no-36-1451 and indicate the wurtzite phase of ZnO. The peak corresponding to (002) indicates the traces of MWCNT. The crystallite size of as-synthesized ZnO using Debye-Scherer's equation was 73.8 nm which reduced to 70.16 nm after the formation of the hybrid composite.

$$D = \frac{k\lambda}{\beta_d(\cos\theta)} \tag{1}$$

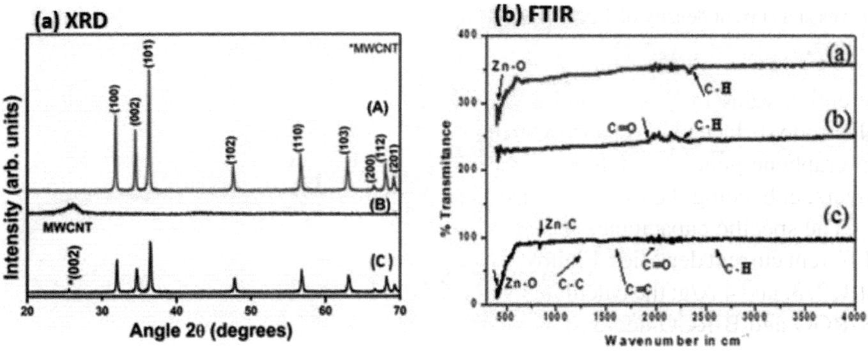

Fig. 4 **a** XRD patterns of ZnO, MWCNT, and ZnO/f-MWCNT composite, **b** FTIR—ZnO, MWCNT, ZnO/f-MWCNT

where 'β' is the full width at half maximum (FWHM) intensity of the peak, 'λ' is X-ray wavelength, and 'θ' is the diffraction angle.

The spectrum shown in Fig. 4b provides surface functional groups of ZnO/f-MWCNT nanocomposite, wherein the peaks below 500 cm^{-1} indicate Zn–O stretch, 1112 cm^{-1} indicate C–C stretch. The peaks at 1345, 1442, and 1570 cm^{-1} indicate the C=C stretching of COOH group. The peaks at higher frequencies 1987, 2192 cm^{-1} indicate C=O group and 2341 cm^{-1} indicate C–H group. The peaks at 3780 cm^{-1} indicate the O–H group.

Surface morphology plays an important role in gas sensing. The morphology characteristic of ZnO/f-MWCNT composite was studied using SEM (Secondary electron microscope).

From Fig. 5, the presence of hexagonal structure of ZnO is connected in the fiber structure of MWCNT as seen in Fig. 5a. A few hollow structure of ZnO is seen in the SEM figure. The addition of functionalized MWCNT provides a porous and rough surface. This morphology is highly beneficial with high aspect ratio and promotes the adsorption of target gas [21, 22].

Energy dispersive X-ray spectroscopy (EDS) study as shown in Fig. 5b shows the presence of Zn, O, and C along with some impurities. Atomic force microscopy (AFM) is used for the surface analysis of the sensor film with a mechanical probe. The thickness and roughness of deposited film was calculated using Gwyddion software as shown in Fig. 6. Thickness is the step height between two peaks and is found to be approx. 560 nm and roughness approximately 120 nm for ZnO/f-MWCNT film. On the other hand, for ZnO film thickness is 150 nm and roughness around 50 nm.

Using UV–Vis spectrophotometer, optical absorbance and energy band gap of ZnO/f-MWCNT composite was studied as shown in Fig. 7. From Fig. 7a, it is observed that composite formation causes a small shift in the absorbance range from 356 to 360 nm indicating a red shift and due to this the band gap energy reduced from 3.0 to 2.97 eV as seen in Fig. 7b.

Energy bandgap in eV was calculated using Tauc's equation (2).

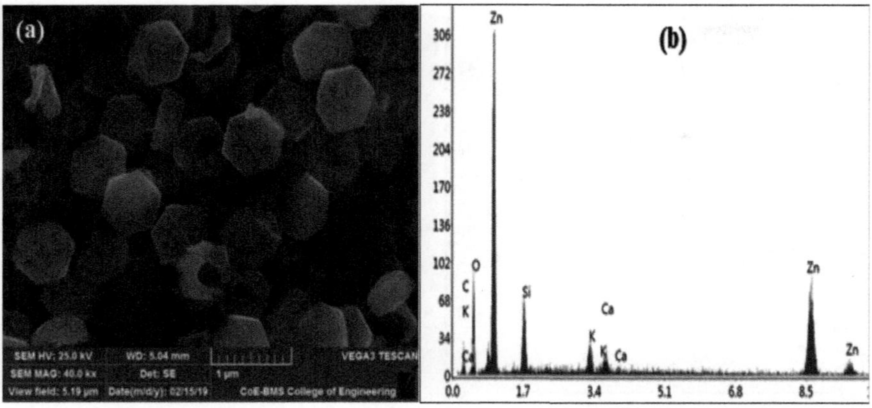

Fig. 5 **a** SEM analysis of ZnO/f-MWCNT, **b** EDS study of ZnO/f-MWCNT

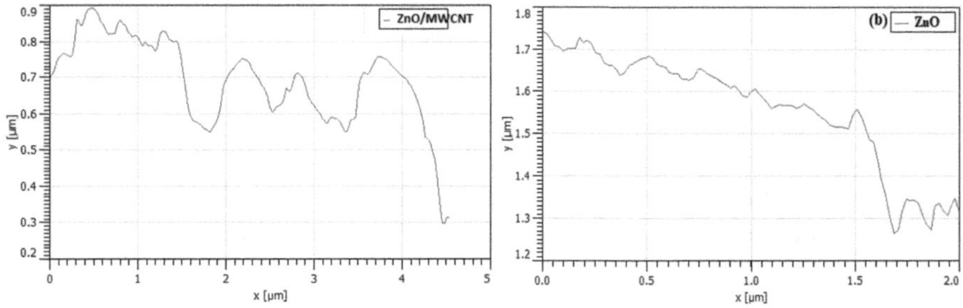

Fig. 6 AFM study of **a** ZnO/f-MWCNT **b** ZnO film showing thickness and roughness

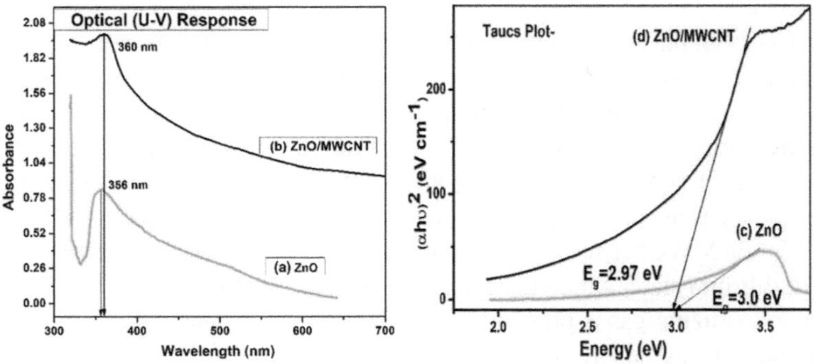

Fig. 7 Optical UV–Vis response **a** ZnO, **b** ZnO/f-MWCNT and Tauc's plot, **c** ZnO, **d** ZnO/f-MWCNT

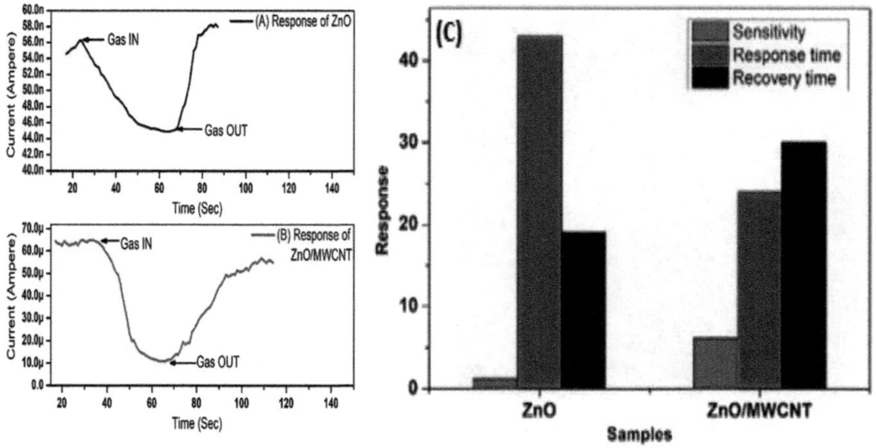

Fig. 8 Gas sensing response of **a** ZnO and **b** ZnO/f-MWCNT sensor, **c** comparative analysis of parameters in terms of response time and recovery time

$$Ahv = A(hv - E_g) \tag{2}$$

where α is absorption coefficient, h is Plank's constant, v is the frequency of incidence light, A is a constant, and E_g is the optical bandgap of the material.

3.2 Gas Sensing Study

The surface boundary of the ZnO/f-MWCNT film is rich with oxygen vacancies that help in the adsorption and desorption process. At room temperature, the sensor did not show any variation in the (conductivity) current, hence the substrate temperature (operating temperature) was increased. At 150 °C variation was observed. When ZnO/f-MWCNT sensor is exposed to air, oxygen molecules are adsorbed onto the sensor surface as negatively charged ions by capturing the free electrons from the n-type of ZnO/f-MWCNT resulting into flow of current [23, 24]. The sensor then goes to an equilibrium state, which is a reference current. When the target gas NO (Nitric Oxide) an oxidizing gas is purged to ZnO/f-MWCNT film sensor, it is adsorbed on the film and during adsorption, it captures one more electron from the conduction band of the film of the sensor. This leads to the chemisorption of NO to NO^- ion and the electron charge carrier in the film reduces, resulting in a decrease in the current, as seen in Fig. 8. The oxidation–reduction phenomena of nitric oxide gas on the surface of the ZnO/f-MWCNT film sensor is shown in Eqs. (3) and (4) respectively.

$$NO_{\text{gas}} + e^- \rightarrow NO^-_{(\text{ads})} \tag{3}$$

$$NO^-_{(ads)} \xrightarrow{\text{reduction}} NO_{gas} + e^- \tag{4}$$

When the NO gas is taken out of the chamber by purging the ambient air into the chamber, the captured electron is released back to the conduction band, thereby carrier concentration increases and the current through the film increases and the sensor reaches back to the reference value; the sensor reaches back to the equilibrium state (Fig. 8). Sensitivity, response and recovery time of sensors are calculated. Sensitivity, response, and recovery time of sensors are calculated and presented in Fig. 8c. Sensitivity of a sensor (S) is calculated as the ratio of current through the sensor in air to current in presence of target gas as given in Eq. 5.

$$S = I_a/I_g \tag{5}$$

where I_a is the current through the sensor in air and I_g is the current through the sensor in presence of target gas. Response time is defined as the time taken by the sensor to reach 90% of the equilibrium value from reference value. Recovery time is defined as the time taken by the sensor to reach 90% of reference value from equilibrium value. From the graph, sensitivity of as prepared ZnO sensor towards NO gas is 1.41, whereas ZnO/MWCNT is 6.3. The exceptional properties of MWCNT like excellent electronic conductivity and relatively high oxidation stability with large specific areas facilitate heterogeneous catalysis and make it more sensitive to the target gas when compared to the as-prepared ZnO thin film sensor.

The response time of composite sensor is 23 s and recovery time is 30 s. On the other hand, response time of ZnO is 43 s and recovery time is 16 s. The response and recovery times of a sensor depend upon the rate of adsorption and desorption of target gas, which depends upon the operating temperature [25]. MWCNT has shown a slow recovery time as compared to ZnO.

A comparative study of various material-based sensors developed to sense NO_x at a particular operating temperature is shown in Table 1. The results are compared with the present work here.

4 Conclusions

The comprehensive analysis of microstructure, optical, and surface properties of ZnO/f-MWCNT composite is carried out using XRD, FTIR, UV–Vis spectrophotometer, SEM, and AFM. Microstructure analysis shows the crystallite size of the composite to be 70.3 nm from XRD. Hollow hexagonal disk of ZnO embedded in the MWCNT fiber of the composite is obtained from SEM which is effective for gas sensing application. The gas sensing studies of the ZnO and ZnO/MWCNT composite sensors towards 2.5 ppm of Nitric oxide showed a sensitivity of 1.41 and 6.1 respectively at an operating temperature of 150 °C. A quick response time was exhibited by the composite sensor due to the high surface-to-volume ratio of

Table 1 Comparative analysis of different materials towards sensing NO_x gas

Materials	Structure	Concentration	Response time (s)	Recovery time (s)	Sensitivity	Operating temperature (°C)	Ref.
ZnO	Nanorod	5 ppm	9	18	4.5	UV source (365 nm LED)	[26]
	Nanoflowers	5 ppm	28	49	1.44		
	Nanosphere	5 ppm	116	215	2.94		
ZnO	Ultra-thin nanoplate	0.5 ppm	500	1000	76	200	[27]
ZnO	Porous nanosheet	2 ppm	500	1000	11.5	200	[28]
ZnO/SWCNT	Nanocomposite	1000 ppm	70	100	9.2	150	[29]
ZnS	Mesoporous shell	1 ppm	100	250	41.12	130	[30]
WO_3	Hollow sphere	300 ppb	670	2940	15.1	RT	[31]
ZnO/MWCNT	Hollow disk structure	2.5 ppm	23	30	6.3	150	Present work
ZnO	Hexagonal disk structure	2.5 ppm	43	16	1.41	150	

MWCNT. MWCNT in the composite enhances current conduction and acts as catalysis, which is evidenced from the current variation in the gas sensing graph. Thus the composite sensor has shown low resistivity, good mechanical robustness, and good integration potential and can be used in NO monitoring at lower ppm level.

Acknowledgements We are thankful to the CENSE lab of IISc, Bangalore, for providing the facilities to carry out gas sensing studies.

References

1. Zhang J, Ghirardo A, Gori A, Albert A, Buegger F, Pace R, Georgii E, Lindermayr C et al (2020) Improving air quality by nitric oxide consumption of climate-resilient trees suitable for urban greening. Front Plant Sci 11:549913. https://doi.org/10.3389/fpls.2020
2. Gomes JBA, Rodrigues JJPC, Rabelo RAL, Kumar N, Kozlov S (2019) IoT-enabled gas sensors: technologies, applications, and opportunities. J Sens Actuator Netw 8:57. https://doi.org/10.3390/jsan8040057
3. Malik R, Tomer VK, Mishra YK, Lin L (2020) Functional gas sensing nanomaterials: a panoramic view. Appl Phys Rev 7:021301. https://doi.org/10.1063/1.5123479
4. Galstyan V, Kaur N, Zappa D, Nunez-Carmona E, Sberveglieri V, Comini E (2020) Chemical gas sensors studied at sensor lab, brescia (Italy): from conventional to energy-efficient and biocompatible composite structures. Sensors (Basel) 20(3):579. https://doi.org/10.3390/s20030579

5. Jana R, Hajra S, Mary Rajaitha P, Mistewicz K, Kim HJ (2022) Recent advances in multifunctional materials for gas sensing applications. J Environ Chem Eng 10(6):108543. https://doi.org/10.1016/jece.2022.108543
6. Van Duy L, Nguyet T, Hung CM, Le Thanh DTN, Hoa ND, Di Natale C (2021) Ultrasensitive NO_2 gas sensing performance of two dimensional ZnO nanomaterials: nanosheets and nanoplates. Ceram Inter 47(20):28811–28820. https://doi.org/10.1016/j.ceramint.2021.07.04
7. Kanan SM, El-Kadri OM, Abu-Yousef IA, Kanan MC (2009) Semiconducting metal oxide-based sensors for selective gas pollutant detection. Sensors (Basel) 9(10):8158–8196. https://doi.org/10.3390/s91008158
8. Filipovic L, Lahlalia A (2018) Review-system-on-chip SMO gas sensor integration in advanced CMOS technology. J Electrochem Soc 165(16):B862. https://doi.org/10.1149/2.0731816jes
9. Odom TW, Huang JL, Kim P, Lieber CM (2020) Structure and electronic properties of carbon nanotubes. J Phys Chem B 104(13):2794–2809. https://doi.org/10.1021/jp993592k
10. Zhang W-D, Zhang W-H (2009) Carbon nanotubes as active components for gas sensors. J Sens 160698. https://doi.org/10.1155/2009/160698
11. Ayman LH, Naumovski N, Maher WA, Ghanem A (2017) Application of carbon nanotubes in chiral and achiral separations of pharmaceuticals, biologics and chemicals. Nanomater (Basel) 7(7):186. PMID: 28718832. https://doi.org/10.3390/nano7070186
12. Srikantaswamy S (2015) Dispersion of multiwall carbon nanotubes in organic solvents through hydrothermal supercritical condition. J Nanomater Article ID 381275. https://doi.org/10.1155/2015/381275
13. Sabri FNAM, Zakaria MR, Akil HM (2020) Dispersion and stability of multiwalled carbon nanotubes (MWCNTs) in different solvents. AIP Conf Proc 2267:020043. https://doi.org/10.1063/5.0024711
14. Marcel G, Luciano W, Renan R, João S, Benjamim C (2020) Improving the dispersion of MWCNT and MMT in PVDF melts employing controlled extensional flows. J Appl Polym Sci 138(17):50274. https://doi.org/10.1002/app.50274
15. Dilonardo E, Penza M, Alvisi M, Rossi R, Cassano G, Di Franco C, Palmisano F, Torsi L, Cioffi N (2017) Gas sensing properties of MWCNT layers electrochemically decorated with Au and Pd nanoparticles. Beilstein J Nanotechnol 8:592–603. https://doi.org/10.3762/bjnano.8.64
16. Díaz-Corona N, Javier M, Jose P-L, David de la Luz A, Maria R, Robles-Águila M (2019) Structural, optical and electrical behavior of zinc oxide/MWCNT composite thin films. Opt Quant Electron J 51. https://doi.org/10.1007/s11082-019-1935-4
17. Rahul P, Vinay G, Vikas S, Jitendra B, Ashok S, Ajay A (2018) Effect of IDE placement on response in metal oxide gas. Mater Res Exp 9:096415. https://doi.org/10.1088/2053-1591/aad728
18. Vinoth E, Gopalakrishnan N (2020) Fabrication of interdigitated electrode based on ZnO sensors for room temperature ammonia detection. J Alloy Comp 824:153900. https://doi.org/10.1016/j.jallcom.2020.153900
19. Manjunath RV, Sumit S, El Mansouri B, Jia W, Andre B, Duco B, de Smet LCPM, Sudhölter EJR, Zhang GQ (2019) A low-power MEMS IDE capacitor with integrated microhotplate: application as methanol sensor using a metal-organic framework coating as affinity layer. Sensor 19(4):88. https://doi.org/10.3390/s19040888
20. Bindu P, Thomas S (2014) Estimation of lattice strain in ZnO nanoparticles: X-ray peak profile analysis. J Theor Appl Phys 8:123–134. https://doi.org/10.1007/s40094-014-0141-9
21. Chengxiang W, Longwei Y, Luyuan Z, Dong X, Rui G (2010) Metal oxide gas sensors: sensitivity and influencing factors. Sensors 10:2088–2106. https://doi.org/10.3390/s100302088
22. Pargoletti E, Hossain UH, Di Bernardo L, Chen H, Thanh T-P Josh L-D, Cappelletti G, Tricoli A (2019) Room-temperature photodetectors and VOC sensors based on graphene oxide-ZnO nano-heterojunctions. Nanoscale 11:22932. https://doi.org/10.1039/c9nr08901b
23. Dongfang Y, Yan Y, Yang G, Xu JL, Zhang M, Kuo CC, Wang SD (2021) Conducting polymer-inorganic nanocomposite-based gas sensors: a review. Sci Technol Adv Mater 6; 21(1):768–786. https://doi.org/10.1080/14686996.2020.1820845

24. Nisreen M, Wasan A-M, Saleh R (2020) Functionalized multi-walled carbon nanotubes network sensor for NO_2 gas selection at room temperature. AIP Conf Proceed 2290:050031. https://doi.org/10.1063/5.0028240
25. Hojati T, Ebrahimi M, Afzalzadeh R (2018) Highly sensitive CO sensor based on ZnO/MWCNT nano sheet network grown via hydrothermal method. Mater Chem Phys 207:50–57. https://doi.org/10.1016/j.matchemphys.2017.12.043
26. Wang H, Dai M, Li Y, Bai J, Liu Y, Li Y, Lu G (2020) The influence of different ZnO nanostructures on NO_2 sensing performance. Sens Actuators B: Chem 329:129145. https://doi.org/10.1016/j.snb.2020.129145
27. Van Duy L, Nguyet TT, Hung CM, Le DTT, Van Duy N, Hoa ND, Biasioli F, Tonezzer M, Di Natale C (2021) Ultrasensitive NO_2 gas sensing performance of two dimensional ZnO nanomaterials: nanosheets and nanoplates. Ceram Int 47:28811–28820. https://doi.org/10.1016/j.ceramint.2021.07.042
28. Choi MS, Kim MY, Mirzaei A, Kim H-S, Kim S-i, Baek S-H, Chun DW, Jin C, Lee KH (2021) Selective, sensitive, and stable NO_2 gas sensor based on porous ZnO nanosheets. Appl Surf Sens 568:150910. https://doi.org/10.1016/j.apsusc.2021.150910
29. Barthwal S, Singh B, Nakshatra BS (2018) ZnO-SWCNT nanocomposite as NO gas sensor. Mater Today: Proceed 5(7):15439–15444. https://doi.org/10.1016/j.matpr.2018.05.030
30. Li Y, Shan L-X, Lian X-X, Zhou Q-J, Min D (2021) Enhanced NO_2 sensing performance of ZnO@ZnS core-shell structure fabricated using a solution chemical method. Ceram Int 47:27411–27419. https://doi.org/10.1016/j.ceramint.2021.06.163
31. Zhao J, Hu M, Liang Y, Li Q, Zhang X, Wang Z (2020) A room temperature sub-ppm NO_2 gas sensor based on WO_3 hollow spheres. New J Chem 44:5064–5070. https://doi.org/10.1016/j.apsusc.2021.150910

Development of MWCNT/Co_3O_4 Nanocomposite-Based Electrodes for Electrochemical Sensing of Ascorbic Acid

R. Ranjitha, Aarti S. Bhatt, and C. R. Ravikumar

1 Introduction

Ascorbic acid (AA), often known as vitamin C, is a vital bioactive substance that has a significant role to play in various metabolic processes including free radical scavenging [1], biosynthesis of neurotransmitters [2], collagen [3], and Fe reducing activity [4]. AA is not synthesized by our body, and this water-soluble vitamin is required to be introduced to the body through diet [5]. AA is a labile compound and is subjected to degradation in contact with enzymes and atmospheric oxygen. The oxidation is pronounced due to the activity of metal cations, light, and thermal interference [6]. AA is widely used as an antioxidant for preventing color and flavor change in food/beverage industries [7]. Various pharmaceutical formulations include the addition of AA in order to treat common cold [8], psychiatric and neurodegenerative diseases [9], quenching the reactive oxygen species generated in the body [10], etc. The importance of AA in biological systems makes it worth quantifying at low levels in real samples. Hence, it is the need of the hour to develop well-accurate, quick, and reliable methods for the qualitative and quantitative determination of AA. The difficulty in the identification of AA in the real samples includes its facile oxidation to dehydroascorbic acid and to 2,3-diketo-L-gulonic acid, which is likely to be an irreversible reaction [11]. The irreversibility of the reaction makes the determination of AA difficult. In spite of this, there are many established methods for the real-time analysis and quantification of AA. These methods include colorimetric methods [12], flow injection analysis [13], high-pressure liquid chromatography

R. Ranjitha · A. S. Bhatt (✉)
Department of Chemistry, Affiliated to Nitte (Deemed to be University), NMAM Institute of Technology, Nitte 574110, India
e-mail: aartis@nitte.edu.in

C. R. Ravikumar
Research Centre, Department of Chemistry, East West Institute of Technology, Bengaluru 560091, India

S. T. Aruna et al. (eds.), *Recent Advances in Electrochemical Science and Technology*, Springer Proceedings in Materials 47, https://doi.org/10.1007/978-981-97-3958-5_8

methods [14], and fluorometric determination method [15]. Yet, all these traditional methods suffer from disadvantages like reliability, repeatability, and sensitivity. In recent days, vibrant research is being carried to develop smart electrode materials for the rapid assessment of AA. The electrode-based electrochemical techniques offer a cost-effective, quick, and reliable alternative for the detection of such bioactive materials. Various literatures are available on AA determination by electrochemical techniques with modified electrodes like screen printed melanin-like nanoparticles (NP) modified electrodes [16], modified alginate/CuO [17], CuO-graphene nanocomposites (NC), etc. [18]. Carbon paste electrodes (CPE) modified with additives like NPs, carbon nanotubes (CNT), organic moieties are now more commonly used as sensors because of ease in the fabrication processes, easy renewability, and more significantly, its compatibility with numerous modifiers added to enhance the electrochemical activity [19]. Recently, various metal oxide, polymer-based/metal/metal oxide-based NCs are used for AA sensing via electrochemical measurements. Sun [20] developed AuNP@MoS_2 NC enrooted via electrodeposition for detection of AA along with other analytes. In addition, CNTs, also referred to as molecular-scale wires, are popular for their enhanced mechanical strength and electrical conductivity. The CNTs have been the most captivating material as they offer unique mechanical, electrical, structural, and electrochemical properties inherent within them [21]. The adduct of these CNTs, i.e., single-walled CNTs (SWCNT) or multi-walled (MWCNT) with metal/metal oxide makes them a promising candidate for the sensing studies [22–24]. For the detection of xanthine, Dalkiran et al. [25] reported a chitosan/MWCNT/Co_3O_4 modified glassy carbon electrode biosensor. Anodic and cathodic detections of xanthine were reported to have linear working ranges of 2.0×10^{-7} M to 1.6×10^{-5} M and 2.0×10^{-5} M to 3.6×10^{-4} M, respectively, with detection limits of 2.0×10^{-7} M and 2.0×10^{-5} M, respectively [25].

In the present study, for the determination of AA, the CPE has been modified with adducts of Co_3O_4 nanoparticles (NPs) and Li-doped Co_3O_4 NPs with MWCNTs. To investigate the electrochemical changes caused by the addition of MWCNTs, the metal oxide nanostructures were mixed with MWCNT at various wt%. The working electrodes were then subjected to electrochemical measurements by varying the scan rates (10–50 mV s^{-1}). These modified CPE electrodes were determined for their sensing ability toward bioactive AA. Also, to evaluate their inherent capacitance, modified electrodes were subjected to cyclic voltammetry (CV), electrochemical impedance spectroscopy (EIS), and amperometry studies.

2 Materials and Methods

2.1 Materials

The experiment utilizes MWCNT, which was purchased from the Shilpent brand, with the purity of 99% and about 4–8 layers, with the diameter of 5–20 nm. The length as mentioned is 10 microns and has a black powder appearance. All the solvents and chemicals utilized in the current experiment were of the analytical grade and were bought from Sigma Aldrich, India. Unless otherwise stated, they were utilized directly without any additional purification.

2.2 Preparation of MWCNT/Co_3O_4 NCs

The pristine (MCP) and 1 wt% Li-doped (MC1) Co_3O_4 NPs were prepared via microwave techniques, supplemented with polyethylene glycol-200 (PEG-200) surfactant as reported in our previously published work [26]. In brief, aqueous solution of cobalt acetate was precipitated with NaOH in presence of PEG, which was later subjected to microwave heating (210 W, 5 min, 95 °C). The resulting precursor was alternately washed with water and ethanol and subjected to annealing for 3 h at 350 °C. To prepare the doped sample, lithium citrate was added prior to microwave-treatment. For further synthesis of MWCNT/Co_3O_4 CPE, Co_3O_4 NPs were finely ground with varied weight % (2.5 and 5.0%) of MWCNTs using a pestle and mortar until a homogenized mixture was formed and no solvent was utilized for this purpose. Figure 1 shows the specifications of the sample codes.

2.3 Structural Characterization of MWCNT/Co_3O_4 NCs

The Co-O stretching/bending frequency peak was identified as a component of a preliminary analysis for the proper blending of the MWCNT with the Co_3O_4 nanostructures using a Fourier-transform infrared (FT-IR) spectrophotometer (Shimadzu). Powder XRD (p-XRD) techniques were used to examine the samples for crystal phase purity on a Rigaku miniflex 600 (5th gen) with CuKα radiation (40 kV, 15 mA) at a scan rate of 2° min^{-1}. A field emission scanning electron microscope (FESEM) was used to examine the surface morphology (FESEM; Carl Zeiss, USA), with EHT 5.00 kV and width range of 1.9 mm. A carbon tape was utilized as a base for the analysis of nanocomposites.

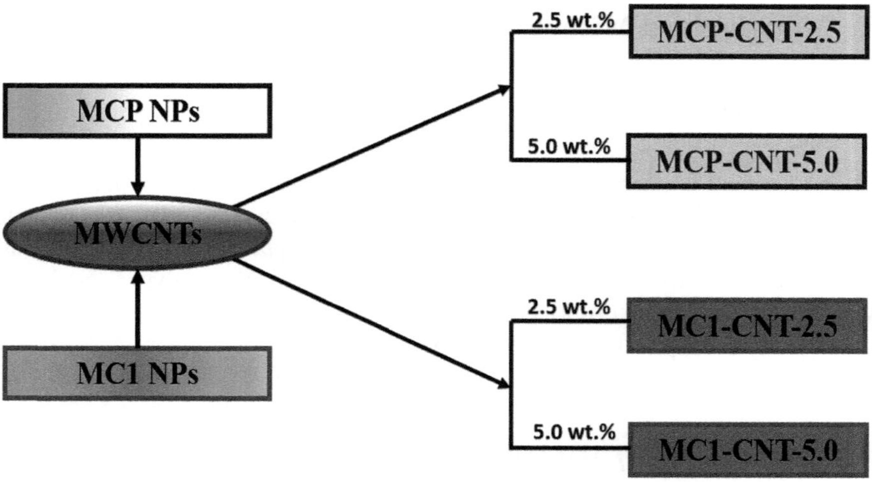

Fig. 1 Flowchart representing preparation of MWCNT/Co_3O_4 nanocomposites (NCs)

2.4 *Electrochemical Investigations and the Fabrication of CPE Modified with MWCNT/MCP and MWCNT/MC1 Electrodes for AA Sensing*

The CV and EIS analyses for the MWCNT/Co_3O_4 NCs were carried out using a CHI604E potentiostat. The electrochemical evaluation was carried out using a three-electrode system with a Pt wire serving as the counter electrode and an Ag/AgCl electrode serving as the reference electrode. The electrolyte utilized was 0.1 M KOH. To prepare the CPE working electrode, 70% graphite powder was mechanically mixed with 15% binder (silicone oil) and MWCNT/Co_3O_4 NCs were used to make a carbon slurry. The prepared paste was inserted into a specially designed Teflon rod with a 3 mm dent and an internal copper wire to create the electrochemical connections. Sandpaper was employed to smooth out the paste. The electrochemical studies, viz., CV and amperometric measurements were performed with scan rates ranging from 10 to 50 mV s^{-1}. For AA sensing, the bioactive substance was given a 30-min sonication in aqueous media, before suspending in the electrolyte solution, in varying concentrations of 1–6 mM.

3 Results and Discussions

3.1 FT-IR Spectroscopy Studies

Figure 2a depicts FT-IR spectra of pristine MWCNT. The absorption band at 1552 and 1630 cm^{-1} are due to the C=C stretching of the hexagonal or benzenoid structure of the MWCNT [27, 28]. The vibration bands at 1300 and 1732 cm^{-1} are attributed to –CH bending and –C=O stretching modes, respectively [29]. Figure 2b portrays FT-IR spectra of pristine Co_3O_4 NPs. The Co_3O_4 NPs show two sharp peaks around 654.68 and 551.92 cm^{-1}. These peaks correspond to the stretching frequency of tetrahedral Co^{+3} and octahedral Co^{+2} in Co_3O_4 while a low intense absorption band corresponding to 1641.29 and 1357.04 cm^{-1} is due to O–H bending modes of adsorbed water species during the synthesis [26]. Figure 2c shows the FT-IR spectra of MWCNT/Co_3O_4 NCs. The absorption bands at 1555 and 1308 cm^{-1} correspond to weak C=C and –C=O stretching which shows the proper blending of MWCNT with Co_3O_4 nanostructures. The band at 828 cm^{-1} corresponds to –CH_2 bending vibrations of surfactant PEG-200.

3.2 p-XRD Analysis

Figure 3a displays the p-XRD diffraction pattern of MWCNT while Fig. 3b depicts the p-XRD patterns of pristine and Li-doped Co_3O_4 nano samples which are in agreement with JCPDS card no. 43-1003, reflecting the formation of Co_3O_4 structure [26]. However, on addition of MWCNT additive, a strong diffraction peak at $2\theta = 26.2°$, corresponding to the Miller indices (002) plane is observed (Fig. 3c). This peak is in agreement with JCPDS card no. 01-0646 corresponding to MWCNT [30]. The co-existence of the diffraction peaks corresponding to Co_3O_4 NPs along with the diffraction peak of MWCNT reveals the perfect blending of MWCNT with Co_3O_4 NPs. Also, the absence of shift in peak position validates the nanocomposite formation.

3.3 Electron Microscopy Measurements

The details of the surface features of the modified electrodes at various composite compositions are derived from their electron microscopy features. Figure 4 depicts the FESEM images of MWCNT/Co_3O_4 NCs at different magnifications (1 μm, 200 nm, and 100 nm). The FESEM micrographs of pure MWCNT show tubular structure. However, images of 2.5 and 5% MWCNTs with MCP/MC1 exhibit flake-like agglomerated morphology thereby indicating the modification of the surface of the electrodes. Also, visual observation of images makes it clear, that an increase in

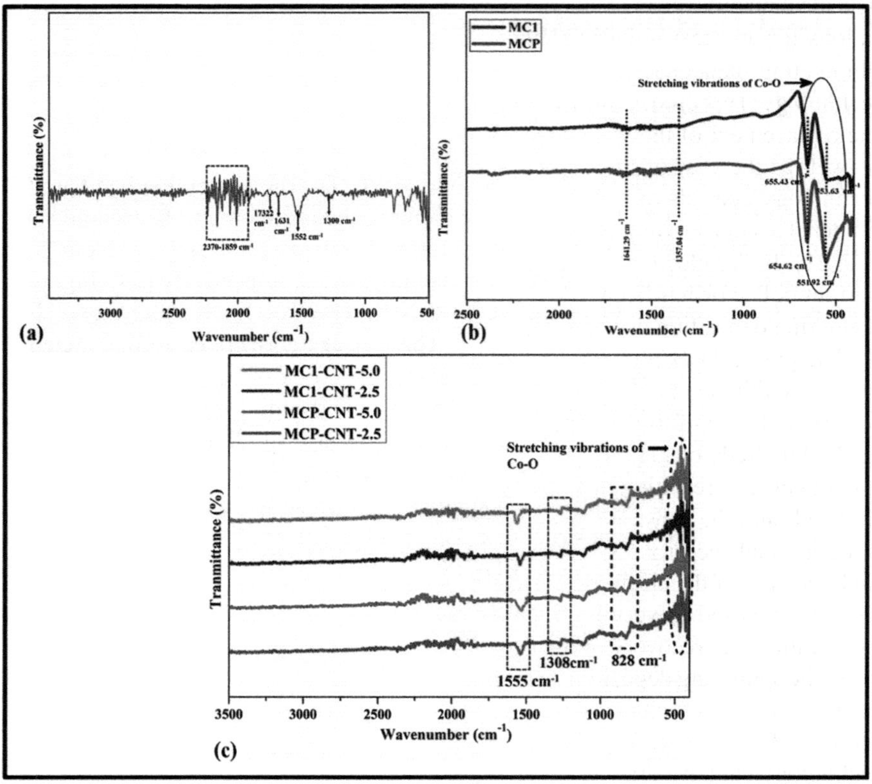

Fig. 2 FT-IR spectra of **a** pristine MWCNT; **b** pristine Co_3O_4 NPs; and **c** Co_3O_4-MWCNTs-based NCs

the percentage of MWCNT increases the agglomeration of doped composites. Similar results have been reported by Sourabh Kumar and Benedict [31] for MWCNT-TiO_2 nanocomposites. The authors have attributed the detangling of the MWCNT tubular structure on their association with TiO_2 NPs.

3.4 Electrochemical Studies

Figure 5a–d represents the cyclic voltammograms of MWCNT/Co_3O_4-based electrodes recorded at various scan rates varying from 10 to 50 mV s^{-1}. An increase in the scan rate from 10 to 50 mV s^{-1} pronounces current responses with detectable anodic peaks, which arise due to redox reaction that occurs at the electrodes surface due to the electron transfer process [32]. The resultant nanocomposites of MWCNTs along with MCP and MC1 NPs at different wt%, reveal cyclic voltammograms not nearing rectangular behavior. The oxidation and reduction peaks are thought to be

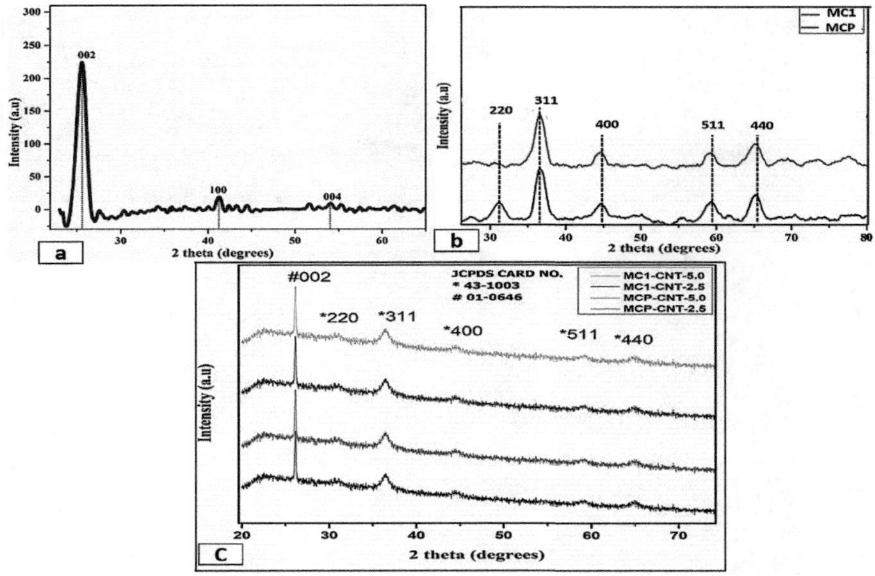

Fig. 3 p-XRD pattern of **a** Pristine MWCNT, **b** Co_3O_4 NPs and **c** Co_3O_4-MWCNTs-based NCs

influenced by pseudo-capacitance, which is a product of the electrochemical process [33]. The CV graph shows that as the scan rate increases, the redox peaks rise as well. On increasing the scan rate, the anodic peaks also shift to a positive potential value.

Further, the Randles–Sevcik equation (Eq. 1) was used to gain a better understanding of the redox process and species transport on the electrode surface.

$$i_p = \left(2.69 \times 10^5\right) n^{3/2} \times AD^{1/2} \times C_o \times v^{1/2} \tag{1}$$

where

- n stoichiometric number of electrons involved in the electrode reaction
- A electrode's area (cm^2)
- D diffusion coefficient of the electro active species ($cm^2\ s^{-1}$)
- Co concentration of electroactive species ($mol\ cm^{-3}$)
- v scan rate ($mV\ s^{-1}$)
- i_p maximum current (A).

The concentration of constructed bioelectrode can be determined using Eq. 2 as shown below

$$C_0 = \frac{\rho}{M} \tag{2}$$

where

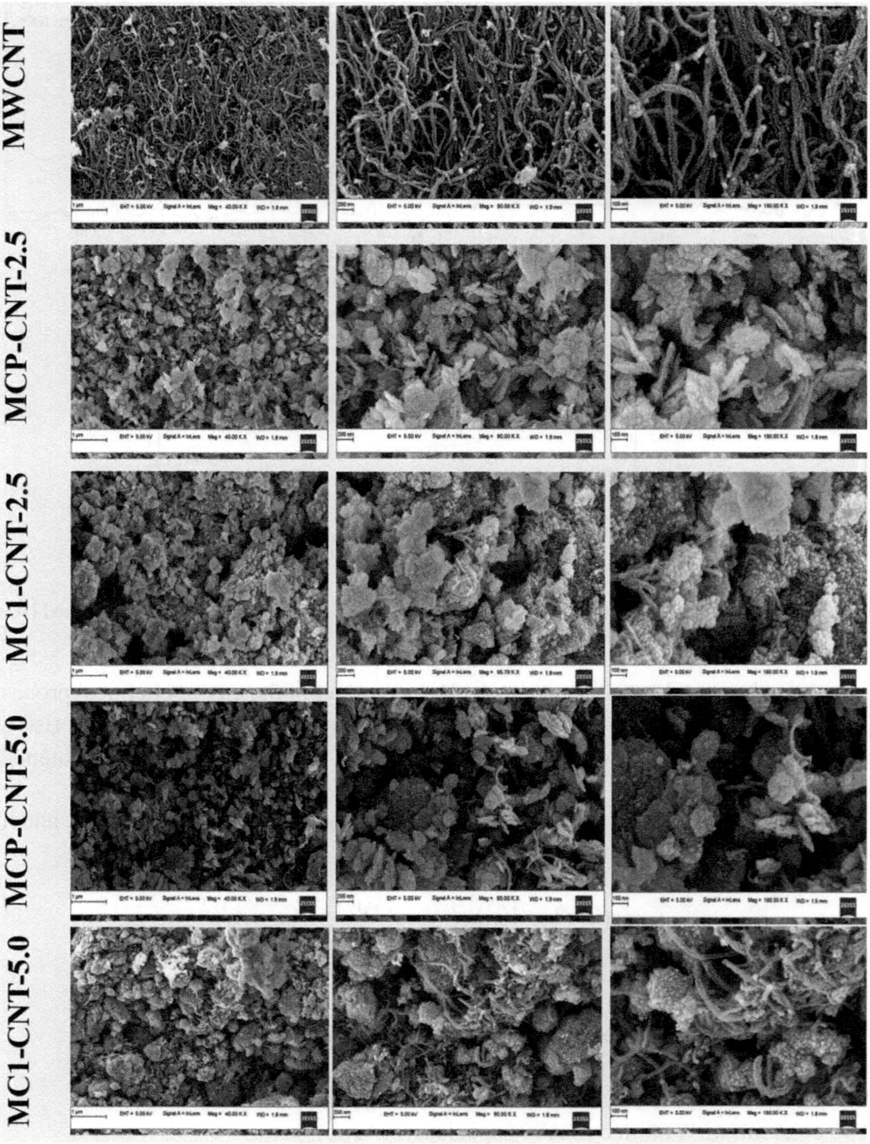

Fig. 4 FESEM images of MWCNT/Co_3O_4 NCs at different magnifications

ρ theoretical density and M = molar mass of the electroactive material.

Figure 5e shows the linearly fitted peak current response versus square root of scan rate. It supports the concept of diffusion-controlled electron transport [34, 35]. The slope of the straight lines gives diffusion coefficient (D) for the NC electrode, the values of which are given in Table 1. The MC1-CNT-5.0 has the best D value

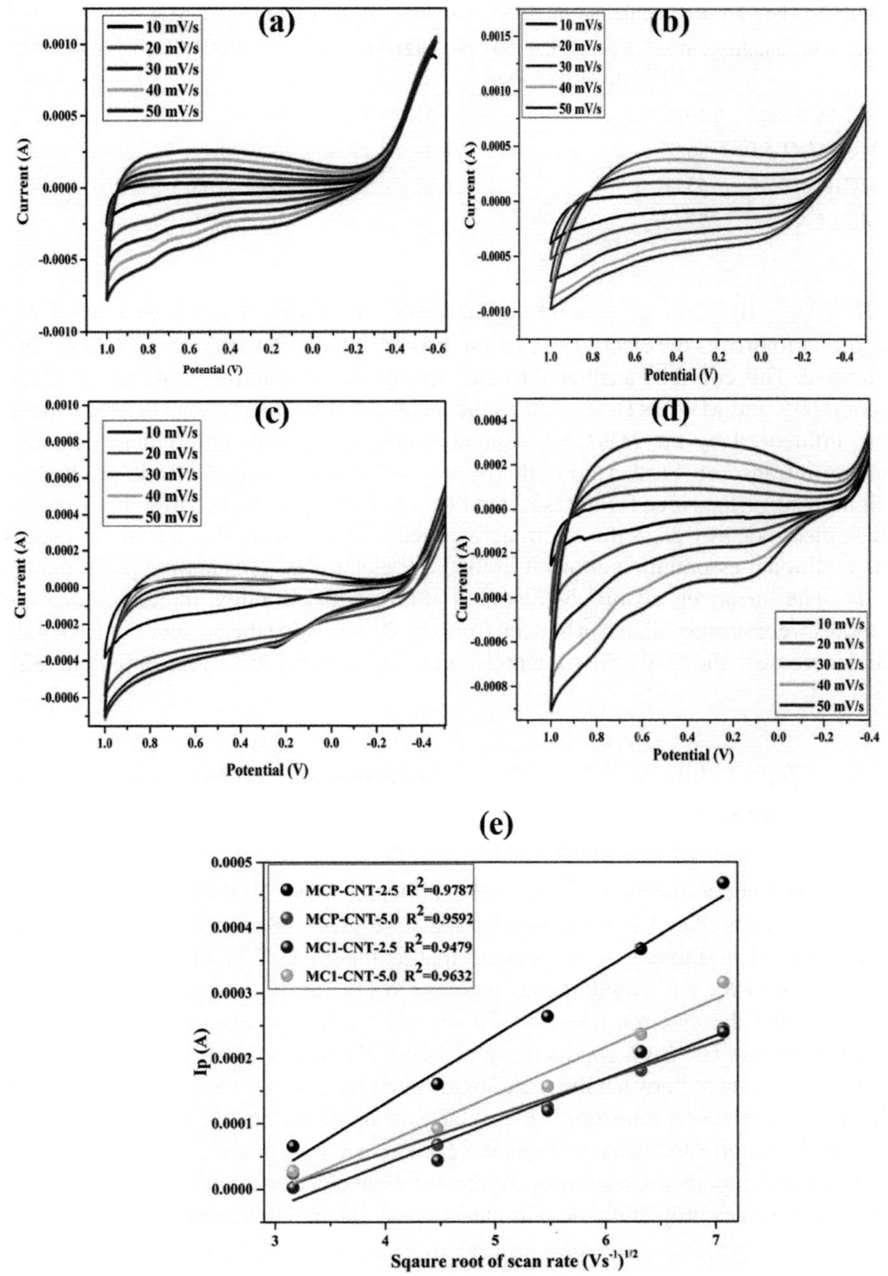

Fig. 5 Cyclic voltammograms at different scan rates for **a** MCP-CNT-2.5, **b** MCP-CNT-5.0, **c** MC1-CNT-2.5, **d** MC1-CNT-5.0, **e** plot of anodic peak current response versus square root of scan rate

Table 1 The diffusion coefficient and EIS circuit fitted values for MC-CNT NCs

Name of sample	Diffusion coefficient (D) $\times 10^{-4}$ ($cm^2\ s^{-1}$)	R_{ct} (Ω)	C_{dl} (F)	LOD $\times 10^{-3}$ (mol L^{-1})
MCP-CNT-2.5	10.03	25.55	3.978×10^{-8}	1.65
MCP-CNT-5.0	56.05	18.44	2.894×10^{-7}	0.99
MC1-CNT-2.5	65.06	16.95	5.596×10^{-3}	0.63
MC1-CNT-5.0	73.05	16.31	6.469×10^{-3}	0.55

of $73.05 \times 10^{-4}\ cm^2\ s^{-1}$ among the fabricated electrodes. The 5.0 wt% MWCNT addition improves the electrode reaction process compared to the 2.5 wt% MWCNT additive. This could be attributed to the fact that the conductive properties of metal oxide NPs and MWCNTs, as well as the weak covalent interactions between them, are influenced by the MWCNT content leading to considerable enhancement in electron transport. Furthermore, the porous MWCNT on the electrode produces a significant surface area for unrestricted flow of electrolytes and charges between the base electrode as well as the electroactive species at the electrode's surface, which is an additional essential component in the CPE electrodes' outstanding performance [36]. The linear regression coefficient is almost equal to unity, indicating that the studies were successful. It can be seen from the D values that the presence of Li dopant and increase in the MWCNT content (by wt%) enhance the diffusion coefficient [32].

3.5 Electrochemical Impedance Spectroscopy (EIS) Measurements

Figure 6a, b shows the Nyquist plot of the fabricated MWCNT-based electrodes. The EIS parameters like charge transfer resistance (R_{ct}), double layer capacitance (C_{dl}) are tabulated in Table 1. It is observed that as the wt% of MWCNT is increased, a decrease in R_{ct} and simultaneous increase in C_{dl} can be observed which clearly exhibits that the electron transport for the MC1-CNT-5.0 electrode is faster than for the others. The high conductivity of MWCNT makes it an effective electron-conducting source between the metal oxide and electrode surface, a property which favors quick electron transfer [37]. Similarly, in another literature, CNT-based PANI electrodes manifested enhanced specific capacitance than pristine PANI sheets [38]. Several such reports are testimony for the excellent electrochemical activity of CNT-based electrodes projecting them as capacitance-enhancing materials.

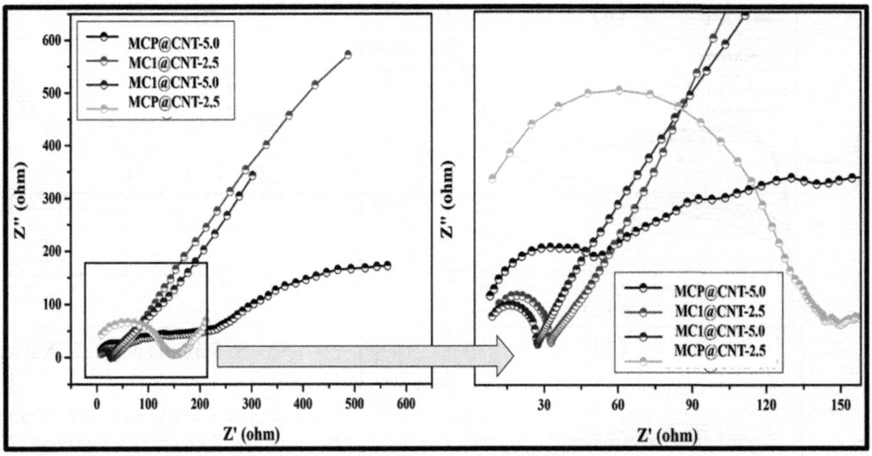

Fig. 6 Nyquist plot of **a** MWCNT-based pristine/doped NCs, **b** enlarged view

3.6 Ascorbic Acid Sensing by Fabricated MCP-CNT and MC1-CNT Electrodes

The efficacy of the surface-modified MCP-CNT and MC1-CNT electrodes for the detection of AA is tested. The cyclic voltammograms of MCP-CNT and MC1-CNT for the detection of AA with concentrations varying from 1 to 6 mM are shown in Fig. 7a–d. The figure shows that redox peak increases with the concentration of AA, indicating that the constructed electrode is sensitive to AA concentration. As a result, MWCNT-modified metal oxide-based electrodes have a lot of potential for sensor development for bioactive chemicals like AA.

From Fig. 5, the oxidation peaks observed at scan rate of 30 mV s^{-1} are 0.139 V (for MCP-CNT-2.5); 0.048 V (for MCP-CNT-5.0); 0.138 V (for MC1-CNT-2.5); and 0.243 V (for MC1-CNT-5.0). However, the cathodic peaks remain undetectable. Now, a careful analysis of cyclic voltammograms on AA sensing at concentration range of 1–6 mM (Fig. 7a–d) shows that the anodic peak occurs at 0.184 V (for MCP-2.5); 0.211 V (for MCP-5.0); 0.253 V (for MC1-2.5); and 0.270 V (for MC1-5.0). It was noted that with the addition of AA, the peaks experience a further shift (positive shift) thereby indicating the oxidation of AA. Literature shows a similar shift in the redox peaks on sensing the paracetamol drug by bio-fabricated green α-Fe_2O_3 NPs [39]. At larger concentrations of AA, the mass transfer phenomenon causes a change in the redox peaks, which is balanced by higher potential application [40]. The plot of I_p (A) versus AA concentration (mM) in Fig. 7d has an average regression coefficient (R^2) of 0.9623. For AA concentrations in the range of 1 mM to 6 Mm, the R^2 is approaching unity which clearly validates that the electrochemical process proceeds via diffusion-controlled phenomena [41]. Further, an attempt has been made to assess the LOD value using Eq. 3 and the results are shown in Table 1. MC1-CNT-5.0 showed the best detection limit of 0.55×10^{-3} mol L^{-1}, which could

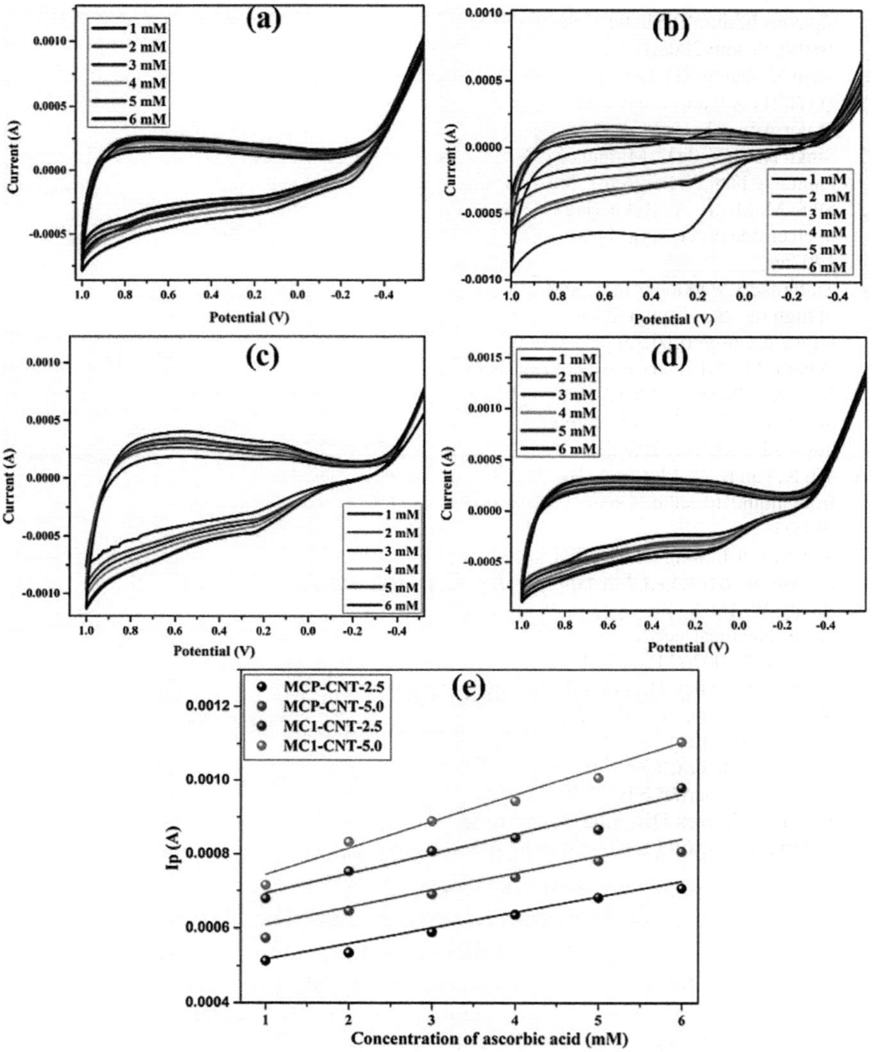

Fig. 7 Cyclic voltammograms on AA sensing at concentration range of 1–6 mM **a** MCP-CNT-2.5, **b** MCP-CNT-5.0, **c** MN1-CNT-2.5, **d** MN1-CNT-5.0, **e** anodic–cathodic peak current response versus concentration (1–6 mM) of AA

be explained by its significant characterization results.

$$LOD = 3 * \left(\frac{SD}{N}\right) \quad (3)$$

where N and SD are respectively the slope and standard deviation of AA calibration plot.

Table 2 Comparison of the developed sensor for AA sensing with previously reported sensors in literature

Electrode	LOD (μM)	Ref.
Cobalt(II)tetrasulfophthalocyanine-/GCE	45.17	[43]
$NiCoO_2$/C modified GCE	0.5	[44]
Polymer (e-MIP) SPE	20	[45]
Au/ RGO/ GCE	51	[46]
MC1-CNT 5/CPE	550	This work

The sensing mechanism of metal oxides-based sensors depends generally on its non-stoichiometric nature, metal/oxygen vacancies, and oxygen deficiencies inside the lattice. These parameters are greatly influenced by the synthetic method [42]. Also, the intrinsic properties of metal oxides can be tuned using different dopants. The dopants have the capacity to modify the electronic structure of the material which may result in increased carrier mobility, variation in conductivity, reduction in the band gap. The doped sample (MC1) had a lower band gap (3.07 eV) than the undoped parent material (MCP) (3.21 eV) resulting in a better electrochemical performance [26]. A comparison of the LOD values indicates that the inclusion of Li in the Co_3O_4 structure increases the electrode sensitivity, which is attributable to its improved conductivity. Table 2 provides a comparison of the limit of detection with those of previously reported sensors for AA.

3.7 *Amperometry Measurements*

Chronoamperometry is a time-dependent technique where a square-wave potential is applied to the working electrode. The current of the electrode, measured as a function of time, fluctuates according to the diffusion of an analyte from the bulk solution toward the sensor surface. Chronoamperometry can therefore be used to measure current–time dependence for the diffusion-controlled process occurring at an electrode. These changes depend on the analyte concentration [47].

The chronoamperometric plots of fabricated CNT-based electrodes for sensing AA at a given concentration range from 1 to 6 mM are given in Fig. 8. The sweep range of potential values chosen is –1.4 to 0.2 V for the analysis. The current response was recorded for each successive addition of AA. A time gap of 50 s was maintained, between successive analyte addition of 1 mM. An enhancement in current was observed, which thereafter reaches the steady-state after a definite period. Such a trend in the current values reveals the quick response toward sensing of AA by the fabricated electrode [39, 48].

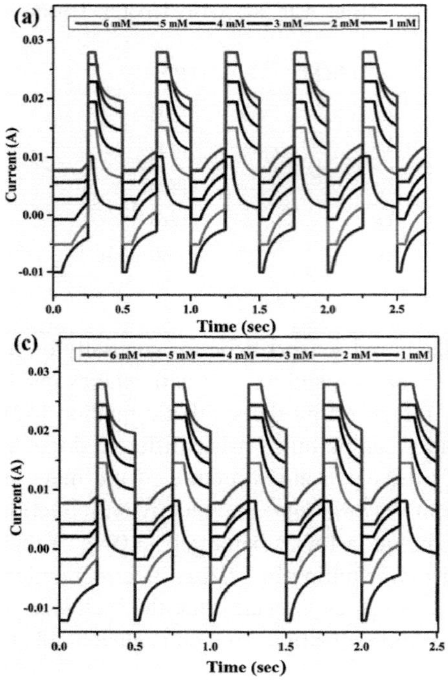

Fig. 8 Current versus time plots on AA sensing at concentration range of 1–6 mM **a** MCP-CNT-2.5, **b** MCP-CNT-5.0, **c** MC1-CNT-5.0, **d** MC1-CNT-5.0

4 Conclusions

Pristine and Li-doped Co_3O_4 NCs with MWCNT at different wt% (2.5 and 5.0) has been successfully fabricated. The prepared NCs were characterized via FT-IR, p-XRD, and FESEM studies. The fabricated nanocomposites were checked for preliminary blending of MWCNT and nano samples via FT-IR spectroscopy. The diffraction peaks of MWCNT and cobalt oxide observed in the p-XRD diffraction pattern confirmed the formation of the nanocomposite. The microscopic images revealed a change in morphology after the addition of MWCNT to Co_3O_4 NPs. The inclusion of the MWCNT further improved the electron transfer process. A preliminary investigation on the sensor studies was carried out through CV and chronoamperometry. The results revealed a quick and instant oxidation of ascorbic acid leading to an increased current response. Further, the LOD measurement showed the best results for composites fabricated with Li-doped metal oxides. Thus, the results of CNT-based electrodes for electrochemical research are more promising and suitable for development of amperometric sensors.

References

1. Siddeeg SE, Alsaiari N, Tahoon M, Ben Rebah F (2020) The application of nano-materials as electrode modifiers for the electrochemical detection of ascorbic acid: review. Int J Electrochem Sci 15:3327–3346. https://doi.org/10.20964/2020.04.13
2. Harrison F, May J (2009) Vitamin C function in the brain: vital role of the ascorbate transporter (SVCT2). Free Radic Biol Med 46:719–730. https://doi.org/10.1016/j.freeradbiomed.2008.12.018
3. Darr D, Combs S, Pinnell S (1993) Ascorbic acid and collagen synthesis: rethinking a role for lipid peroxidation. Arch Biochem Biophys 307:331–335. https://doi.org/10.1006/abbi.1993.1596
4. Smirnoff N (2018) Ascorbic acid metabolism and functions: a comparison of plants and mammals. Free Radic Biol Med 122:116–129. https://doi.org/10.1016/j.freeradbiomed.2018.03.033
5. Nimse SB, Pal D (2015) Free radicals, natural antioxidants, and their reaction mechanisms. RSC Adv 5:27986–28006. https://doi.org/10.1039/C4RA13315C
6. Bhagvan NV (2001) Medical biochemistry. e-book. ISBN: 9780080511399
7. Varvara M, Bozzo G, Celano G, Disanto C, Pagliarone CN, Celano CV (2016) The use of ascorbic acid as a food additive: technical-legal issues, (in Eng). Ital J Food Saf. 5:14313. https://doi.org/10.4081/ijfs.2016.4313
8. Bucher A, White N (2016) Vitamin C in the prevention and treatment of the common cold. Am J Lifestyle Med 10:181–183. https://doi.org/10.1177/1559827616629092
9. Kocot J, Luchowska D, Kiełczykowska MI, Kurzepa J (2017) Does Vitamin C influence neurodegenerative diseases and psychiatric disorders? Nutrients 7:659. https://doi.org/10.3390/nu9070659
10. Guaiquil V, Golde D, Beckles D, Mascareno E, Siddiqui M (2004) Vitamin C inhibits hypoxia-induced damage and apoptotic signaling pathways in cardiomyocytes and is-chemic hearts. Free Radic Biol Med 37:1419–1429. https://doi.org/10.1016/j.freeradbiomed.2004.06.041
11. Paixão T, Bertotti M (2008) FIA determination of ascorbic acid at low potential using a ruthenium oxide hexacyanoferrate modified carbon electrode. Journal of pharmaceutical and biomedical analysis.46:528–533. https://doi.org/10.1016/j.jpba.2007.10.033
12. Kyaw A (1978) A simple colorimetric method for ascorbic acid determination in blood plasma, (in Eng). Clin Chim Acta 86:153–157. https://doi.org/10.1016/0009-8981(78)90128-6
13. Yebra-Biurrun MC (2000) Flow injection determination methods of ascorbic acid (in Eng). 52:367–383. https://doi.org/10.1016/s0039-9140(00)00402-1
14. Nováková L, Solich P, Solichová D (2008) HPLC methods for simultaneous determination of ascorbic and dehydroascorbic acids. TrAC Trends Anal Chem 27:942–958. https://doi.org/10.1016/j.trac.2008.08.006
15. Wu X, Diao Y, Sun C, Yang J, Wang Y, Sun S (2003) Fluorometric determination of ascorbic acid with o-phenylenediamine. 59:95–99. https://doi.org/10.1016/S0039-9140(02)00475-7
16. Argoubi W, Rabti A, Ben SA, Raouafi N (2019) Sensitive detection of ascorbic acid using screen-printed electrodes modified by electroactive melanin-like nanoparticles. RSC Adv 9:37384–37390. https://doi.org/10.1039/C9RA07948C
17. Ibarlucea B, Pérez Roig A, Belyaev D, Baraban L, Cuniberti G (2020) Electrochemical detection of ascorbic acid in artificial sweat using a flexible alginate/CuO-modified electrode (in Eng). Mikrochim Acta 187. https://doi.org/10.1007/s00604-020-04510-5
18. Monsef Z (2015) Simultaneous electrochemical determination of acetaminophen, caffeine and ascorbic acid using a new electrochemical sensor based on CuO-graphene nano-composite. RSC Adv 5. https://doi.org/10.1039/C5RA14138A
19. Svancara I, Vytras K, Barek J, Zima JD (2001) Carbon paste electrodes in modern electroanalysis. Crit Rev Analy Chem 31:311–345. https://doi.org/10.1080/20014091076785
20. Sun H (2014) Gold nanoparticle-decorated MoS_2 nanosheets for simultaneous detection of ascorbic acid, dopamine and uric acid. RSC Adv 52:27625–27629. https://doi.org/10.1039/C4RA04046E

21. Motahary M, Ghoreishi SM, Behpour M, Golestaneh M (2010) Electrochemical de-termination of ascorbic acid at the surface of a graphite electrode modified with multi-walled carbon nanotubes/tetradecyltrimethylammonium bromide. J Appl Electrochem 40:841–847. https://doi.org/10.1007/s10800-009-0067-0
22. Prasad R, Bhat B (2015) Multi-wall carbon nanotube–NiO nanoparticle composite as en-zyme-free electrochemical glucose sensor. Sens Actuators B: Chem 220:81–90. https://doi.org/10.1016/j.snb.2015.05.065
23. Krivetskiy V, Andreev MD, Efitorov AO, Gaskov AM (2021) Statistical shape analysis pre-processing of temperature modulated metal oxide gas sensor response for ma-chine learning improved selectivity of gases detection in real atmospheric conditions. Sens Actuators B: Chem 329:129187. https://doi.org/10.1016/j.snb.2020.129187
24. Singh A, Singh M, Verma N (2020) Electrochemical preparation of Fe_3O_4/MWCNT-polyaniline nanocomposite film for development of urea biosensor and its application in milk sample. J Food Measur Charac 13:1–13. https://doi.org/10.1007/s11694-019-00278-2
25. Dalkiran B, Kaçar C, Erden P, Kiliç E (2014) Amperometric xanthine biosensors based on chitosan-Co_3O_4-multiwall carbon nanotube modified glassy carbon electrode. Sens Actuators B: Chem 200:83–91. https://doi.org/10.1016/j.snb.2014.04.025
26. Ranjitha R, Bhatt AS, Ravikumar CR, Santosh MS, Jayanna CR, Nagaswarupa HP, Sakai K, Madhyastha H (2019) Electroactive Li incorporated cobalt oxide nanostructures for photocatalytic applications. Mater Res Exp 6:1150d6. https://doi.org/10.1088/2053-1591/ab5033
27. Wei XP, Zhang RQ, Wang LB, Luo YL, Xu F, Chen YS (2019) Novel multi-walled carbon nanotubes decorated with gold nanoparticles with poly(2-methacryloyloxyethyl ferrocenecar-boxylate) grafted on to form organic–inorganic nanohy-brids: preparation, characterization, and electrochemical sensing applications. J Mater Chem C 7:119–132. https://doi.org/10.1039/C8TC05294
28. Mahdavi H, Sajedi M, Shahalizade T, Heidari AA (2020) Preparation and application of catalytic polymeric membranes based on PVDF/cobalt nanoparticles supported on MWCNTs. Polym Bull 77:4489–4505. https://doi.org/10.1007/s00289-019-02983-w
29. Vinodh R, Sangeetha D (2013) Quaternized poly(styrene ethylene butylene poly sty-rene)/multiwalled carbon nanotube composites for alkaline fuel cell applications. Nano-sci Nanotechnol 13(8):5522–5533. https://doi.org/10.1166/jnn.2013.7445
30. Moazzen M, Khaneghah AM, Shariatifar N, Ahmadloo M et al (2019) Multi-walled carbon nanotubes modified with iron oxide and silver nano-particles (MWCNT-Fe_3O_4/Ag) as a novel adsorbent for determining PAEs in carbonated soft drinks using magnetic SPE-GC/MS method. Arab J Chem 12:476–488. https://doi.org/10.1016/j.arabjc.2018.03.003
31. Sourabh Kumar S, Benedict T (2019) Influence of TiO_2 and MWCNT nanoparticles dispersion on microstructure and mechanical properties of Al6061 matrix hybrid nano-composites. Mater Res Exp 6:1265f3. https://doi.org/10.1088/2053-1591/ab5dfe
32. Sandford C, Edwards MA, Klunder KJ, Hickey DP, Li M, Barman K et al (2019) A synthetic chemist's guide to electroanalytical tools for studying reac-tion mechanisms. Chem Sci 10:6404–6422. https://doi.org/10.1039/C9SC01545K
33. Manjunatha AS, Pavithra NS, Shivanna M, Nagaraju G, Ravikumar CR (2020) Synthesis of Citrus Limon mediated SnO_2-WO_3 nanocomposite: applications to photocatalytic activity and electrochemical sensor. J Environ Chem Eng 8:8104500. https://doi.org/10.1016/j.jece.2020.104500
34. Ravi Kumar CR, Santosh MS, Nagaswarupa HP, Prashantha SC, Yallappa, Anil Kumar MR (2017) Synthesis and characterization of β-$Ni(OH)_2$ embedded with MgO and ZnO nanopar-ticles as nanohybrids for energy storage devices. Mater Res Exp 4:065503. https://doi.org/10.1016/j.apsadv.2020.100049
35. Fayemi O, Adekunle A (2015) Metal oxide nanoparticles/multi-walled carbon nanotube nanocomposite modified electrode for the detection of dopamine: comparative electro-chemical study. J Biosens Bioelectr. https://doi.org/10.4172/2155-6210.1000190

36. Zulfli NHM, Bakar AA, Chow WS (2013) Mechanical and water absorption behaviors of carbon nanotube reinforced epoxy/glass fiber laminates. J Reinfor Plast Comp 32:1715–1721. https://doi.org/10.1177/0731684414526
37. Fayemi OS, Adekunle AS, Ebenso EE (2017) Electrochemical determination of serotonin in urine samples based on metal oxide nanoparticles/MWCNT on modified glassy carbon electrode. Sens Bio-Sens Res 13:17–27. https://doi.org/10.1016/j.sbsr.2017.01.005
38. Wang J, Chen Y, Hu Z, Ge Y, Dong G, Hu T, Jhun CG (2020) Preparation and electrochemical properties of functionalized multi-walled carbon nanotubes @ Carbon quantum dots @ Polyaniline ternary composite electrode materials. Appl Sci 10:5462. https://doi.org/10.3390/app10165462
39. Navada KM, Nagaraja GK, D'Souza JN, Kouser S, Ravikumar CR, Manasa DJ (2021) Biofabrication of multifunctional quasi-spherical green α-Fe_2O_3 nanostructures for paracetamol sensing and biomedical applications. Ceram Int 47:33651–33666. https://doi.org/10.1016/j.ceramint.2021.08.275
40. Pisoschi A, Pop AI, Stanca L, Bilteanu L, Serban A (2021) Antioxidant determination with the use of carbon-based electrodes. Chemo Sens 9:72. https://doi.org/10.3390/chemosensors9040072
41. Keerthi M, Boopathy G, Chen SN, Chen TW, Lou BS (2019) A core-shell molybdenum nanoparticles entrapped f-MWCNTs hybrid nanostructured material based non-enzymatic biosensor for electrochemical detection of dopamine neurotransmitter in biological samples. Sci Rep 9:13075. https://doi.org/10.1038/s41598-019-48999-0
42. Sampath Kumar S, Rubio EJ, Noor-A-Alam M, Martinez G, Manandhar S et al (2013) Structure, morphology and optical properties of amorphous and nanocrystalline gallium oxide thin films. J Phy Chem C 117(8):4194. https://doi.org/10.1021/jp311300e
43. Agboola BO, Vilakazi SL, Ozoemena KI (2009) Electrochemistry at cobalt (II) tetrasulfophthalocyanine-multi-walled carbon nanotubes modified glassy carbon electrode: a sensing platform for efficient suppression of ascorbic acid in the presence of epinephrine. J Solid-State Electrochem 13:1367–1379. https://doi.org/10.1007/s10008-008-0691-3
44. Zhang X, Yu S, He W, Uyama H, Xie Q, Zhang L, Yang F (2014) Electrochemical sensor based on carbon-supported $NiCoO_2$ nanoparticles for selective detection of ascorbic acid. Biosens Bioelectron 55:446–451
45. Alberti G, Zanoni C, Magnaghi LR, Biesuz R (2023) Ascorbic acid sensing by molecularly imprinted electrosynthesized polymer (e-MIP) on screen-printed electrodes. Chemosensors 11:348. https://doi.org/10.3390/chemosensors11060348
46. Wang C, Du J, Wang H, Zou C, Jiang F, Yang P, Du Y (2014) A facile electrochemical sensor based on reduced graphene oxide and Au nanoplates modified glassy carbon electrode for simultaneous detection of ascorbic acid, do-pamine and uric acid. Sens Actuators B: Chem 204:302–309. https://doi.org/10.1016/j.snb.2014.07.077
47. Guy OJ, Walker KA (2016) Graphene functionalization for biosensor applications. In: Saddow SE (ed) Silicon carbide biotechnology, 2nd ed. Elsevier, pp 85–141
48. Yang PG, Wang X, Wu LQ, Chen Z, Lin X (2014) Amperometric sensor for ascorbic acid based on a glassy carbon electrode modified with gold-silver bimetallic nanotubes in a chitosan matrix. Microchim Acta 181(1–2):231–238. https://doi.org/10.1007/s00604-013-1104-6

Fig. 1 a Gas sensing ...

2.3 Gas Sensing Set...

The gas sensing studies ... a multimeter Keithley ... substrate holder ... The ... measuring the electrical ... fixed gas ...

... gases were connected ... standard ... through the gas flow ... chamber.

3 Results and Discussion

3.1 Characterization

... pattern of ZnO ... The highest intensity peak ... of 2θ = 36.3° ... match with that of ZnO ... of ZnO. The peak ... crystallite size of ... which reduced to ...

Flexible PS Based ZnO ...

... Sil...

Electrochemical Sensing of Lung Cancer-Related Volatile Organic Compounds Using Polycaprolactone/ Polyaniline Composite Fiber Electrodes

M. S. Sumitha and T. S. Xavier

1 Introduction

Cancer is defined by the presence of uncontrolled cellular proliferation and division, resulting from cellular changes. Apoptosis, or programmed cell death, is an inherent biological process with intrinsic value, despite its outward manifestations. The induction of programmed cell death allows the organism to transition to a more optimized and current state. Malignant cells lack the necessary regulatory elements for controlling their growth arrest and programmed cell death. This causes the buildup of these cells in the body, leading to a lack of oxygen and nutrients for other cells. Malignant cell metastasis can transpire via the lymphatic system [1]. The existence of genetic factors can increase the probability of cancer occurrence. Extended exposure to sunlight and smoking both increase the likelihood of negative health consequences.

Lung cancer exhibits the highest fatality rate in males and stands as the second most prevalent cause of death in females, second only to breast cancer. Lung cancer can be broadly categorized into two main types: Small Cell Lung Cancer (SCLC) and Non-Small Cell Lung Cancer (NSCLC). Non-small cell lung cancer accounts for approximately 85% of all diagnosed cases of lung cancer. Adenocarcinoma, squamous cell carcinoma, and large cell carcinoma are the most common types of non-small cell lung cancer (NSCLC), collectively representing up to 10% of NSCLC tumors [2]. The initial stage of the disease primarily affects the lungs, whereas the second stage is characterized by the tumor being located near the lymph nodes. Stage 3 neoplasm affects the pulmonary system and lymphatic nodes. Stage 3 comprises two subgroups: 3A, which indicates cancer that has only metastasized to one side of the chest, and 3B, which indicates lymph nodes on both sides. The fourth stage of cancer involves metastasis, the spread of cancerous cells from the tumor to other organs [2].

M. S. Sumitha · T. S. Xavier (✉)
Centre for Advanced Materials Research, Department of Physics, Government College for Women, University of Kerala, Thiruvananthapuram 695014, India
e-mail: xavierkattukulam@gmail.com

S. T. Aruna et al. (eds.), *Recent Advances in Electrochemical Science and Technology*, Springer Proceedings in Materials 47, https://doi.org/10.1007/978-981-97-3958-5_9

The identification of early-stage lung cancer (stage 1 or 2) is rarely accomplished due to the presence of nonspecific symptoms like fatigue and coughing, which patients usually attribute to other causes.

Japanese case-controlled studies in the early 2000s showed that chest radiography and sputum cytology can lower lung cancer mortality [3]. Nonspecific symptoms like weariness and coughing might lead patients to misdiagnose stage 1 or 2 lung cancer. Early lung cancer detection reduces patient mortality. This situation requires biosensors that can identify early signs. Multiple methods can detect primary and metastatic lung cancer cells. Modern immunological and molecular methods can discover and study cancer patients' blood circulating tumor cells (CTCs). Circulating tumor cells (CTCs) have malignant characteristics. Lung cancer diagnostics often uses immune magnetic enrichment, immune cytochemical detection, morphology-based enrichment followed by reverse transcriptase polymerase chain reaction (RT-PCR), and RT-PCR [4]. Figure 1 depicts the classification of various CTCs detection techniques. Detecting lung cancer is hard. Image-guided transthoracic needle biopsy may injure patients. Liquid biopsy can detect and analyze cancer because it is minimally invasive. Extracellular vesicle (EV) surface proteins indicate lung cancer. Lei et al. [5] devised a sensitive and specific fluorescence technique. This approach uses padlock probes and ERCA (exponential rolling circle amplification) assay. The study demonstrated that identifying extracellular vesicles linked to lung cancer is very accurate and sensitive, with a detection threshold of 4.222×10^4 particles/mL. Pulmonary nodules and micro-calcifications are frequently categorized as spherical opaque forms with parenchymal lacerations. Diameters under 3 cm are used to evaluate these structures [6]. Computer Tomography (CT) is a prominent nodule detection and diagnosis method due to its sensitivity. Space resolution, cost-effectiveness, prevalence, and non-invasiveness are advantages of the preceding approach.

Due to its three-dimensional chest cavity imaging, computed tomography (CT) is effective at detecting pulmonary nodules and micro-calcifications [7]. This imaging

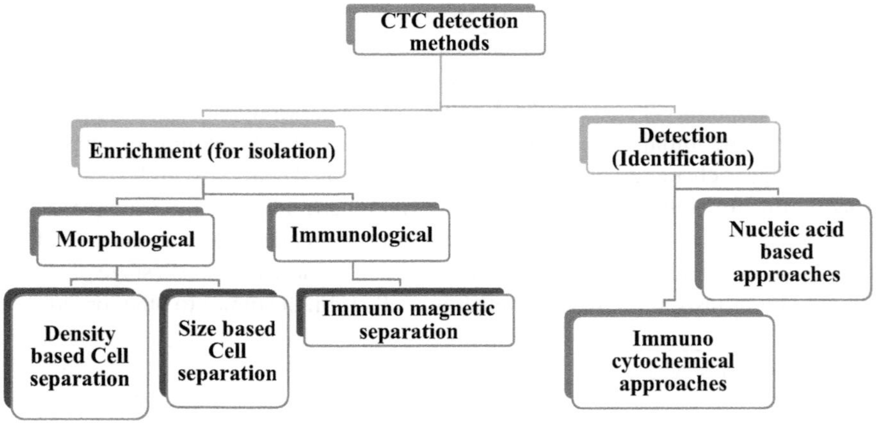

Fig. 1 Various CTCs detection techniques

method yields high resolution nodule, micro-calcification, and pathological tumor pictures. To diagnose patients, radiologists must examine CT scan nodules and micro-calcifications' size, location, texture, and shape. This classic lung cancer screening method is difficult and involves a lot of effort due to misunderstandings and a limited professional understanding of the subject [8]. Total volatile organic compounds (TVOC), aldehydes, benzene, toluene, xylene, etc., are all examples of volatile organic compounds (VOCs) that are involved in photocatalytic reactions in the atmosphere and are not carbon dioxide, carbonic acid, carbonates, metal carbides, or carbonaceous compounds [9]. Despite guidelines for aldehydes and benzene-based organic compounds, scientists worldwide are creating more sensitive, cost-effective, energy-efficient, fast, and reliable VOC sensing technologies [10]. Colorimetric methods only measure organic vapors overall. Although gas chromatography has remarkable separation efficiency, derivatives' low boiling point compromises its determination precision. To compare the various interferences in VOC with the isolated components, liquid chromatography (LC) is difficult. The complex build and high cost of the LC-MS instrument restrict its widespread use. Thus, research on a low-cost, durable, and portable VOC detection system is crucial [10]. Toxicological examination of indoor and outdoor settings, disease diagnosis, and biological processes benefit from accurate and selective VOC detection. Over the past decade, successful VOC sensors and biosensors have been developed and used in many practical applications [11, 12]. Thriumani et al. [13] used GC-MS and SPME to analyze volatile organic chemicals generated by human cells cultured in vitro and relate them to lung cancer indicators. Lung cancer and normal tissue were studied using A549 and W138VA13 cell lines.

The current investigation is predicated on the idea that certain volatile organic compounds (VOCs) from cancer cells can function as the smell signature and be used to give an electrochemical non-invasive lung cancer screening method employing a relatively inexpensive proboscis device. These experiments make use of the selectivity, reproducibility, and sensitivity of nanocomposite electrode materials made from electrospun high molecular weight Polycaprolactone (PCL) and Polyaniline (PANI; in the emerald salt form). It was confirmed using GC-MS tests that the lung cancer cell lines were employed for the electrochemical research.

2 Experimental

Sigma Aldrich supplied the polymers polycaprolactone (PCL) with a molecular weight of 80,000 and polyaniline (PANI) in the emerald salt (ES) form. For the best results when creating a PCL/PANI polymer solution, a mixture of $CHCl_3$ (chloroform) and DMF (N,N-dimethyl formamide) with an optimized $CHCl_3$:DMF solvent ratio of 17:3 was tested. Spectrochem was used to acquire DMF and $CHCl_3$. The above-mentioned solvent ratio was used to prepare 20 ml of an optimized 12 wt% PCL/PANI solution. The optimal concentration of PANI for miscibility testing using 3 mM in 3 ml of DMF was determined. The precursor solutions were stirred for 2 h to ensure the elasticity or viscosity and complete mixing of precursor components which are essential for electrospinning. Electrospinning experimental settings were optimized at a flow rate of 0.5 ml/h, a voltage of 14.2 kV, temperature of 25 °C, 65% relative humidity, and a tip-target distance of 15 cm for the PCL/PANI precursor solution. The electrospun samples were collected from a rotating mandrel (300 rpm) wrapped in aluminum foil. The collected polymer mats were carefully attached to an oxide free nickel foam so as to get an active electrode area of (1×1 cm^2). The non-adhered face of the nickel foam is safely covered with a leak-free insulation tape.

The A549 cell line, which is derived from human alveolar cell carcinoma, was obtained from the National Centre for Cell Sciences (NCCS) located in Pune, India. The technique of microscopy known as inverted phase contrast microscopy is being discussed. After acquiring the cell line, it underwent observation through inverted phase contrast microscopy using a Labomed TCM-400 equipped with a MICAPS™ HD camera from the United States. The topic of discussion pertains to the cultivation of cells in vitro, specifically the composition and upkeep of the culture media.

The cells were cultured in Dulbecco's Modified Eagles Medium (DMEM-Himedia), which was supplemented with 10% heat-inactivated Fetal Bovine Serum (FBS) and 1% antibiotic cocktail comprising Penicillin (100 U/ml), Streptomycin (100 μg/ml), and Amphotericin B (2.5 μg/ml) and followed the standard cell culture procedure. The cells were cryopreserved at a low passage number using liquid nitrogen vapor phase. The modified cell culture media was supplemented with 20% FBS and either 10% DMSO or glycerol. For the preliminary studies 25 cm^2 TC flask samples were used and the optimization and reproducibility checks were done using samples from 75 cm^2 TC flasks. The sample flasks were purchased from Athmic Biotech Solutions Pvt. Ltd., Thiruvananthapuram.

The electrochemical experimental apparatus was assembled using borosilicate glassware, air-tight stoppers, connectors, corks, insulation or cling-on tapes, and a balloon as a secondary protective layer between the culture tube and stopper glassware. The 500-ml borosilicate glass round-bottom flask with three necks held the electrolyte. The electrolyte was 1 M potassium chloride (KCl—purchased from Sigma Aldrich). Three-electrode systems were used for electrochemical experiments. The study used a PCL/PANI electrospun fiber mat affixed to a treated nickel foam, a platinum (Pt) wire electrode, and an Ag/AgCl electrode as working, counter, and reference electrodes. Immersing three electrodes in the electrolyte required inserting them through the flask necks and taking the connection wires outside through the air-sealed corks before connecting them to the cell culture flask tube. The system included a Y-shaped connector and a borosilicate jar with a cap to remove gases from the cell culture flask without any leakage. Figure 2 shows a side-profile model with a protruding trunk.

At room temperature, a Bruker Kappa Apex II was used to record X-ray diffraction (XRD) patterns in the 2θ range of 10–80° with a step size of 0.0167° and a scan speed of 2° min^{-1}. The XRD pattern was recorded using CuK$_\alpha$ radiation at a wavelength of 1.5406 nm. Using an ESCALAB 250XL, Thermoscientific electron spectrometer, X-ray photoelectron spectra (XPS) data were collected. The FTIR spectra were acquired

Fig. 2 Experimental setup

using a Shimadzu IR Prestige 21 spectrometer. Scanning electron microscopy (SEM) using a 5 kV Carl Zeiss EVO 18 Research instrument was used to characterize the surface morphologies. Lateral SEM images of the samples were also obtained using the instrument. Brunauer–Emmett–Teller (BET) surface area analyzer Model: Nova Touch 1 × 4 was used for BET analysis of the electrode materials. The Shimadzu GC-MS (Model Number: QP2010S) was utilized to conduct an analysis of the volatile organic compounds (VOCs) released by cancer cell lines through the process of Gas Chromatography-Mass Spectroscopy. A column with a length of 30 m, an inner diameter of 0.25 mm, and a thickness of 0.25 μm, specifically the ELITE-5MS column, was utilized. The GC-MS Solutions software was utilized in conjunction with the NIST 11 and WILEY 8 libraries.

Using a BioLogic VSP300 electrochemical workstation, we performed cyclic voltammetry (CV) and electrochemical impedance spectroscopy (EIS). Scanning at 200, 150, 100, 50, 20, and 10 mV/s, CV curves were plotted. From the CV curve, the specific capacitance in F/kg of the electrodes were determined using Eq. (1), also expressed [14].

$$C_s = \int IdV/(2mv\Delta V) \quad (1)$$

The area under the CV curve is denoted by $\int IdV$, the potential window by ΔV (in volts), the potential sweep rate by v (in millivolts per second), and the mass of the active material by m (in kilograms). The reaction mechanism at the electrode contact is illuminated by EIS measurements taken between 1 MHz and 0.1 Hz. The cells were subjected to varying perturbation voltages, with amplitudes ranging from 0 to 10 mV.

3 Results and Discussion

Figure 3 portrays the X-ray diffraction (XRD) patterns of the electrode material that has been meticulously selected and optimized. The electrospun PCL/PANI nanocomposite electrode material exhibits distinct identification planes. The (110) and (200) crystallographic planes of PCL have been identified and labeled. The recorded peaks at 21.3° and 23.7° suggest that the reflections originated from an orthorhombic unit cell with crystal parameters similar to polyethylene [15, 16] [JCPDS-00-062-1286]. The peaks located at 15.9°, 20°, 21.9°, and 44.6° are attributed to polyaniline (PANI) and have been designated as (031), (113), (020), and (222) based on the indexing provided by JCPDS-053-1890 [16–21].

The identification of functional groups can be effectively accomplished through the utilization of Fourier Transform Infrared (FTIR) analysis. The Fourier Transform Infrared (FTIR) spectra presented in Fig. 4 exhibit the distinctive peaks corresponding

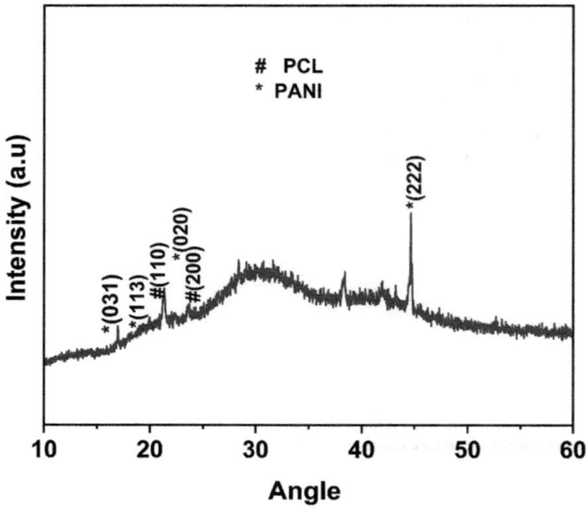

Fig. 3 XRD pattern of electrospun PCL/PANI nanocomposite electrode material

Fig. 4 FTIR spectrum of PCL/PANI nanocomposite

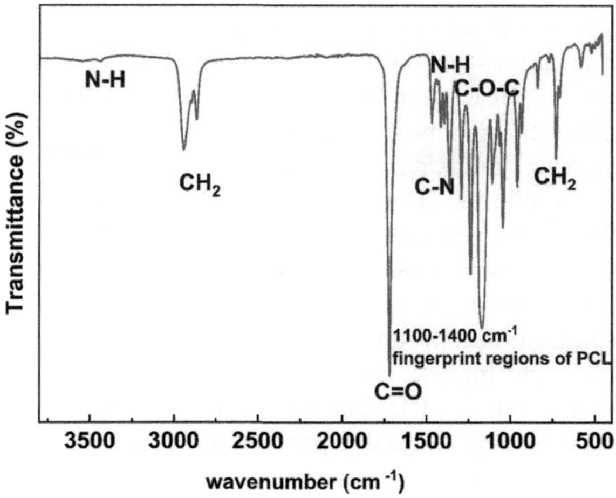

to the constituent bonds of the designated PCL/PANI nanocomposite electrode materials [16]. The infrared spectrum displays the presence of ester carbonyl bonds at a wavenumber of 1728 cm^{-1}. Additionally, the existence of hydroxyl groups is indicated by two peaks observed at wavenumbers of 3540 and 3439 cm^{-1}. The identification of –CH aliphatic stretching bonds at 2945 and 2866 cm^{-1} has been reported [22]. The presence of PCL has been confirmed through FTIR analysis, which has revealed the fingerprint vibrations of PCL in the range of 1100–1400 cm^{-1}. Additionally, the C=O stretching at 1726 cm^{-1}, in conjunction with the C–O–C stretching vibrations at 1042, 1107, and 1233 cm^{-1}, have been observed. These findings align with previous research [16, 23, 24]. The presence of functional groups in PANI [19, 21, 25] can be inferred from the vibrations observed in the C-N stretching region of 1192–1516 cm^{-1}, N–H stretching regions of 1590–1650 cm^{-1}, and 3300–3500 cm^{-1}.

Figure 5 illustrates the X-ray photoelectron spectroscopy (XPS) analysis conducted to examine the elemental composition and valency states of the PCL/PANI composites. The X-ray photoelectron spectroscopy (XPS) survey spectrum obtained from the electrode sample (as depicted in Fig. 5a) indicates the presence of the highest counts per second for binding energy within the range of 280–290 electron volts (eV). The spectral data exhibits distinct peaks in the energy ranges of 530–535 eV and 396–410 eV, which are indicative of transitions involving the C1s, N1s, and O1s orbitals. The identification of the N1s peak is challenging due to the limited energy resolution during the analysis process. This is particularly true when the peak is in close proximity to the C1s peak, and the degree of obscurity is further compounded when the elemental concentration of nitrogen in the sample is low. The deconvolution fitting procedure employed the Savitzky–Golay smoothing technique in conjunction with Gaussian function tools. Figure 5b–d depicts the high resolution X-ray photoelectron spectroscopy (XPS) spectra in the C1s, N1s, and O1s regions, respectively. These spectra serve to confirm the diverse binding states of the aforementioned elements present in the sample. Typically, the sp carbon/carbide bond

[26] is indicated by the peak contribution observed at 282.9 eV in the C1s spectrum. The carbon–carbon bond is designated in the spectral analysis. The peaks of C1s that have undergone deconvolution, exhibiting binding energies of 284.5, 286.1, and 286.9 eV, may be ascribed to carbons present in C–C or C–H, C–N, and C–O or carbonyl group, respectively.

Despite the relatively small quantity of polyaniline (PANI) in comparison to the weight percentage of polycaprolactone (PCL), the presence of a confirmation peak at approximately 286.1 eV provides evidence of a chemical bond between the constituent components within the electrode material. The identification of N1s spectrum deconvolution is attributed to the existence of various chemical species, including Imine/C=N–C (398.5 eV), N^{2+}/N^{3+} (399.9 eV), quaternary ammonium species (401.7 eV), and oxidized N–O species (404.7 eV). These species have been appropriately indexed as per the established literature [16, 21, 27–32].

The residual N components of the electrode substance are fitted and allocated correspondingly [28, 29]. The X-ray photoelectron spectroscopy (XPS) analysis reveals a high resolution spectrum in the O1s region, which has been resolved into two distinct peaks. These peaks have been identified as C–O and H–O–C, respectively.

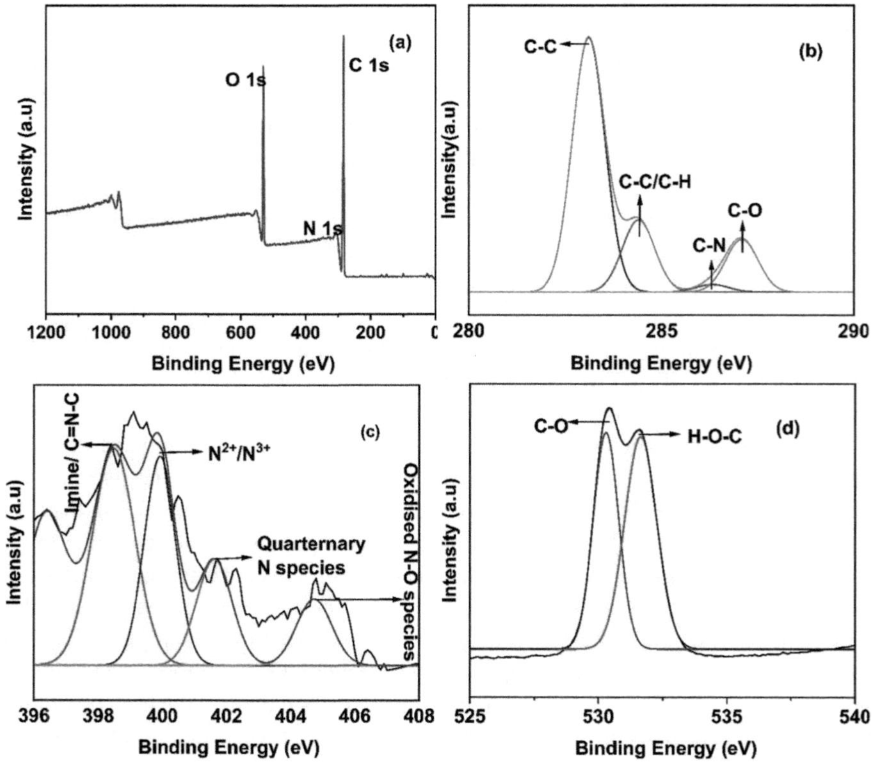

Fig. 5 XPS spectra of PCL/PANI nanocomposite electrode **a** XPS survey scan, **b** C1s region, **c** N1s, **d** O1s

The process of electrospinning is considered to be a versatile and feasible method for producing fibers that are extremely thin. Significant advancements have been achieved in the realm of electrospinning techniques and the customization of electrospun nanofibers to accommodate diverse applications [33]. Additionally, this method can be used to create a variety of fiber assemblies, such as sub-micron springs, convoluted fibers, aligned fiber mesh, patterned fiber mesh, nonwoven fiber mesh, and aligned fiber mesh [34]. The nanocomposite electrode materials, electrospun from PCL and PANI, are exhibited in SEM images in Fig. 6. Nanofibrous nature of the electrodes may be seen clearly in Fig. 6a–c. The shape of the material, which includes porous features, lends credence to its prospective use as a biosensing medium. Fiber diameters are indexed after being measured. The electrode's porous shape increases its active surface area, allowing it to absorb trace amounts of volatile organic gases released by the cell culture flask. The cross-sectional SEM pictures taken at an angle reveal the nanofibers' layered alignment. Electrospun deposition creates holes that easily let molecules pass through to the electrode surface.

BET analysis can determine the surface area and average pore diameter of a material. Figure 7a displays the Nitrogen absorption–desorption isotherms, and Fig. 7b displays the pore size distribution curve obtained by the BJH method of the electrode material. The BET surface area of the sample was found to be 8.918 m^2/g and the average pore radius was calculated to be 2.94 nm. The shape of the nitrogen absorption–desorption isotherms confirms the mesoporous nature of the electrode material.

Images of A549 human lung adenoma cells under an inverted phase contrast microscope, as they grow in flasks, are displayed in Fig. 8. Healthy, mature cells are evident adhering to the flask's surface. It was a well-cultured flask, so after we cut off the nutrients, a lot of the gas the cancer cells make will be trapped inside. The flasks are utilized in GC-MS analysis and electrochemical tests.

Figure 9 depicts the electrochemical behavior of the electrode materials studied by the CV measurements. Both with and without the attached A549 Cell culture flask, cyclic voltammetric readings were taken. The area under the curves for both metrics was shown to grow with increasing scan rates. Although the voltammogram shape is indicative of materials displaying the faradaic process, the optimal potential window range differed significantly between the two cases. Electrode potential windows of 1 to -0.9 V for the electrode that detected gases from A549 cell lines and 0.1 to -1 V for the reference electrode indicated the presence of possible redox reactions. One of the electrodes came into contact with the mysterious gas molecules, whereas the other did not. The biogas detection device connected electrode yielded oxidation and reduction potentials, while the bare Ni foam electrode with the electrospun sample material primarily yielded reduction potential data. The presence of analyte gases can be inferred instantly from the shape and values displayed.

The redox probe exhibits a notable capacity to readily traverse the polymer electrospun porous membrane. The chemical functional groups in the membrane are producing an altered ionic milieu, causing a modification in the effective potential at the surface of the electrode. The term "potential scale" denotes the potential difference existing between the electrode and the reference electrode. When a reactive

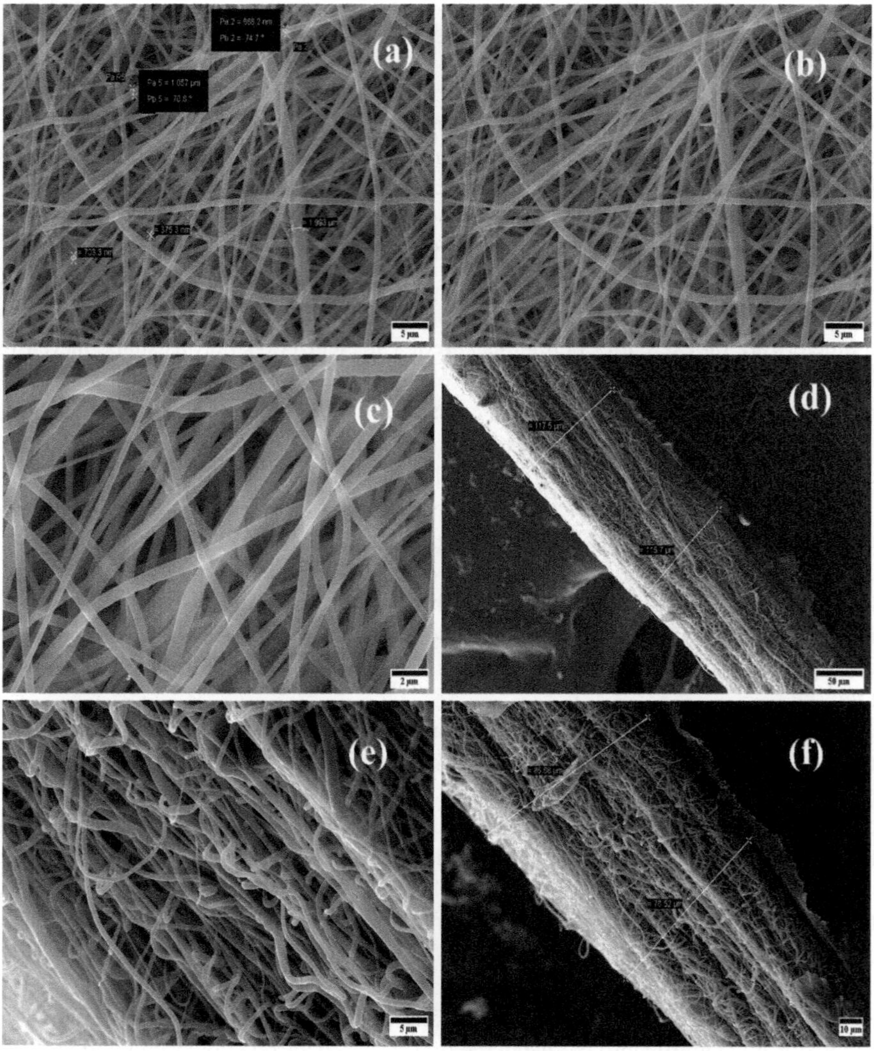

Fig. 6 SEM images of the PCL/PANI nanocomposite electrode material: direct view (**a**, **b**, **c**) and cross-sectional SEM images (**d**, **e**, **f**)

membrane is introduced between the components in detecting the presence of the VOCs, the resulting voltammogram will exhibit a change in the x-scale, thus altering the reference electrode. The obstruction of electrodes, whether partial or complete, would have a substantial impact on reducing the faradaic current. The redox active species is exhibiting a downward shift in potential, suggesting an increasing difficulty in its reduction process. If the role were solely focused on electron transfer, there would be an observed augmentation in peak splitting; which is visible in the anodic peaks in the CV curves of electrospun PCL/PANI electrodes in Ni foam during the

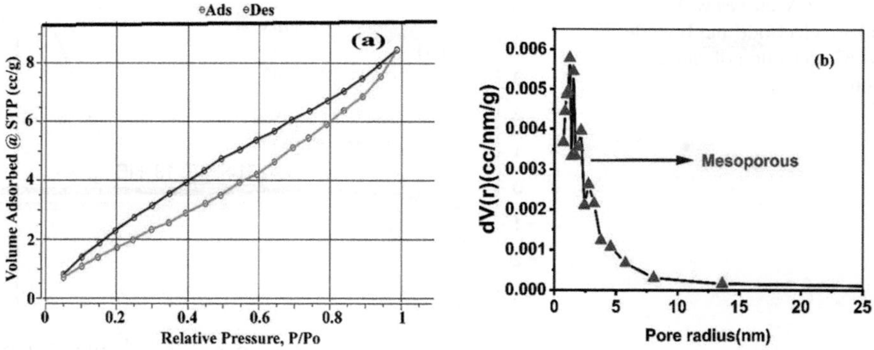

Fig. 7 Nitrogen absorption–desorption isotherms (**a**) and BJH pore size distribution (**b**) of PCL/PANI electrospun nanocomposite electrode material

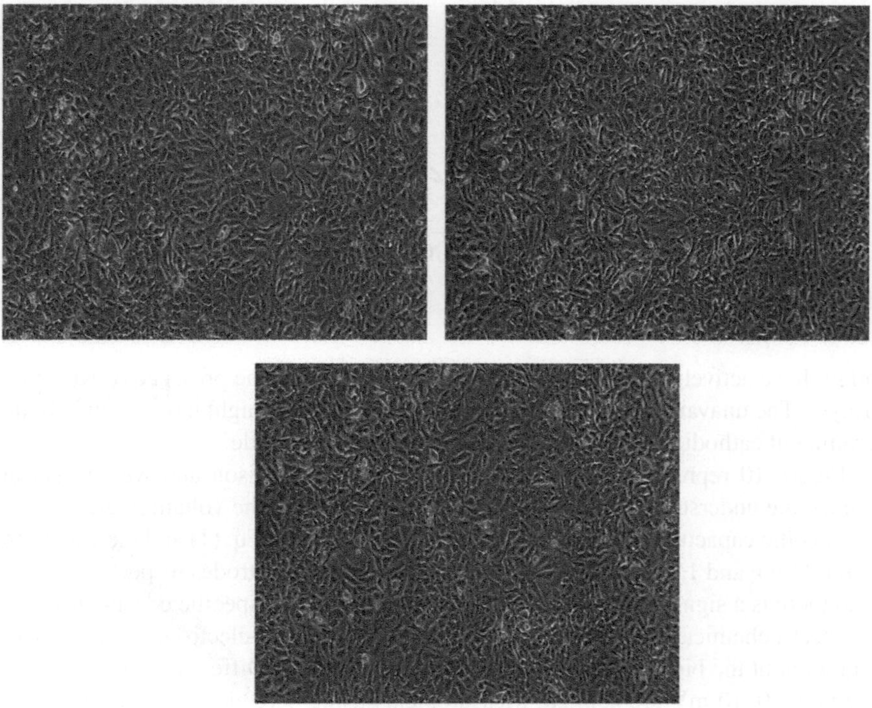

Fig. 8 Inverted phase contrast photographs of A549 cell line

I–V response of the bioanalytes. The organic gases reaching the PCL/PANI nanocomposite electrode in the Ni foam might have penetrated into the electrode surface and

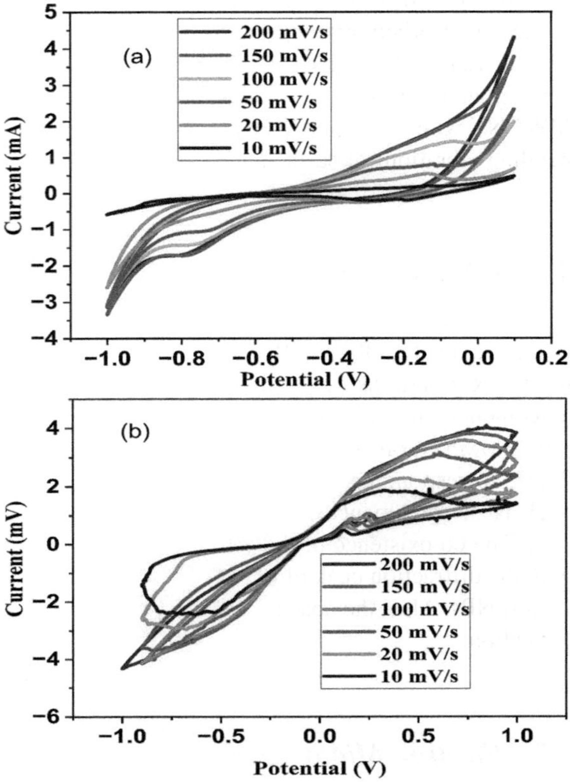

Fig. 9 CV curves with **a** bare electrode and **b** while VOCs detection of cancer cells

might have actively participated in the redox reactions in the presence of KCl electrolyte. The unavailability of these organic gas molecules might have resulted in the prominent cathodic potential in the case of the bare electrode.

Figure 10 represents the lowest scan rate CV comparison and we can get an immediate understanding about the VOCs detection from the voltammogram.

Specific capacitance values are calculated according to Eq. (1) and the values are 2104.9 F/kg and 13,693.5 F/kg for bare and biosample electrodes respectively. Here also there is a significant difference between the values of specific capacitance.

Electrochemical Impedance Spectroscopy (EIS) tests of electrode materials with and without the bioanalyte (VOCs) are shown in Fig. 11. Different perturbations in the range 0–10 mV are applied. Measurements using bare electrodes did not yield a satisfactory electrical response to perturbations greater than 10 mV, whereas the same showed satisfactory electrical responses for higher perturbation potentials also. Ni foam-aided PCL/PANI bare electrode revealed a consistent increasing nature for slopes of the curve at non-zero mV values. In the presence of bioanalyte, the identical electrode failed to produce the same response, particularly at lower frequencies. In the range from 1 MHz to 0.1 Hz, 20 mV of amplitude was selected. A considerable difference in electrode behavior with and without the analyte is seen in the EIS

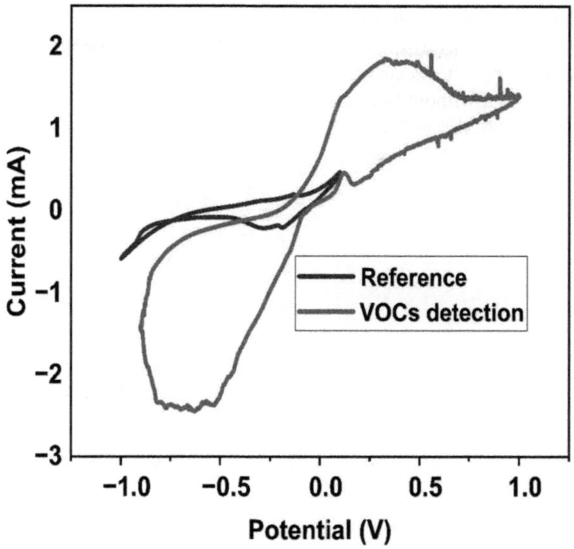

Fig. 10 CV curve comparison at 10 mV/s scan rate

response. In the context of diffusion, it is postulated that the quantity of particles that undergo diffusion over a given cross-sectional area is directly proportional to the concentration gradients present throughout such region. Conversely, in the adsorption process, electroactive species are observed to move from the bulk phase toward the interface. In the case of control study with the bare electrode, the mechanism of diffusion is found to be prominent. The Nyquist plots show that the impedance decreases upon the introduction of VOCs from the cell culture flask to the electrodes.

It was Thriumani et al. [13] who have detected the VOCs from A549 cell lines previously. But they have used GC-MS-SPME, where solid phase microextraction was done and compared the results with normal cell lines. Figure 12 represents the GC-MS data analysis.

Figure 12 presents the gas chromatography-mass spectrometry (GC-MS) data pertaining to the A549 cell line. The figure displays the chromatogram and provides details regarding the identification of peaks. The GC-MS analysis of the sealed cell culture flask reveals the presence of Myristic acid vinyl ester, 2-tetradecyloxirane, stearic acid, vinyl ster, nonadecane, tetradecanal, 3,7,11,15-tetramethyl-2-hexadeceen-1-ol, and glyceryl tristearate. These compounds are detected as specific volatile gases produced by lung cancer cells. There is a lack of specific references available for confirming the identification of volatile organic compounds (VOCs) produced by A549 cells. Table 1 presents the documented volatile organic compounds (VOCs) generated from healthy cells. Nevertheless, the current analytical report of the present investigation lacks the inclusion of any previously identified volatile organic compounds (VOCs) found in normal cells. Based on the available evidence, it can be deduced that the source of origin for the observed entities is likely to be A549 cells. Decane [35, 38–40], nonanal [41–43], tetradecane [38, 44], and other gaseous substances have been identified as potential biomarkers

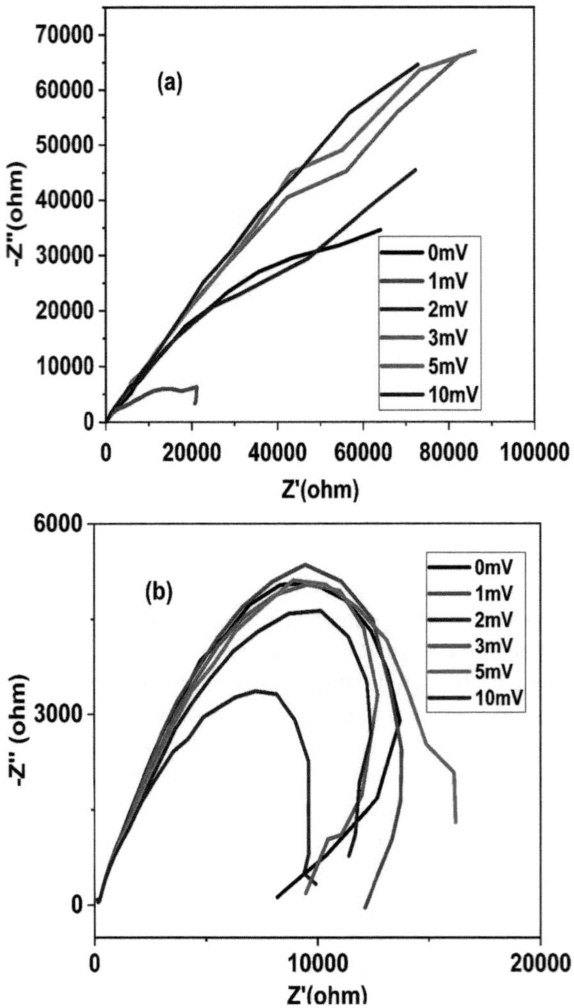

Fig. 11 Nyquist plots of PCL/PANI electrospun nanocomposite electrode material **a** without analyte and **b** with the analyte

for the detection of human lung cancer. The volatile organic compounds (VOCs) depicted in Fig. 12 serve as primary sources for the generation of molecules such as nonadecane, tetradecanal, and others. The verification of VOC detection can be accomplished by utilizing GC-MS data.

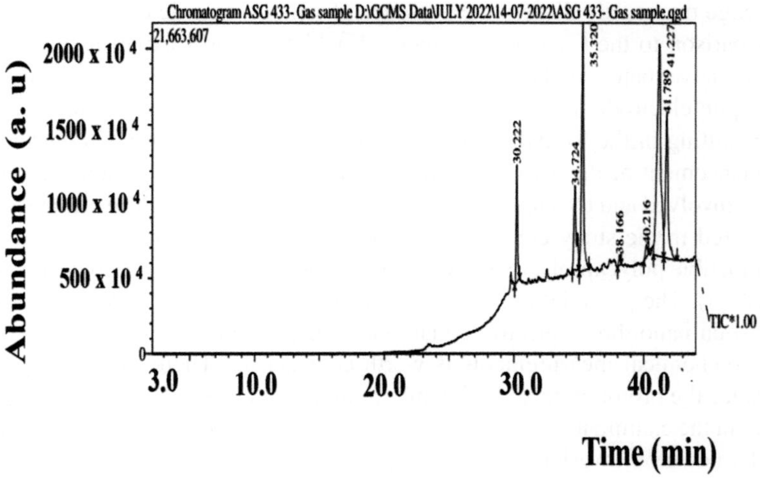

Fig. 12 GC-MS data report

Table 1 VOCs generated by non-cancerous normal cells

	Sl. no	Compound name	Normal cells generating	Reference
Previously reported VOCs	1	1-heptanol	√	[13]
	2	1,2,4-trimethyl benzene	√	[35-37]
	3	Trimethyl[4-(trimethylsilyl) butoxy]silane	√	[13]
	4	2-ethyl-1,4-dimethyl benzene	√	[13]
Reported VOCs in the present study	5	Myristic acid vinyl ester	No reported works confirming the emission of these VOCs by normal cells	
	6	2-tetradecyloxirane		
	7	Stearic acid		
	8	Vinyl ester		
	9	Nonadecane		
	10	Tetradecanal		
	11	3,7,11,15-tetramethyl-2-hexadeceen-1-ol		
	12	Glyceryl tristearate		

4 Summary

The presence of cancer cells is electrochemically identified in this work with the aid of a self-designed, inexpensive, yet efficient electrochemical detection equipment. Even small quantities of volatile organic compounds (VOCs) were sufficient to induce alterations in the form and characteristics of the data obtained from cyclic voltammetry (CV) and electrochemical impedance spectroscopy (EIS) measurements. Given the varying shapes of the CV plots, even those without specialized knowledge may easily discern the presence of volatile organic compounds (VOCs) in comparison to the CT detection method, which requires expert assistance. The detection advantages can be attributed solely to the distinctive characteristics of the electrospun electrode material. The material underwent the process of electrospinning, resulting in the formation of a porous structure. This porosity contributed to the enhancement of the active interaction between the volatile organic compounds (VOCs) involved and the electrochemical detecting mechanism. The electrode material utilized in the study consisted of a composite electrospun mat comprising of biocompatible polycaprolactone (PCL) and a conducting polymer known as polyaniline (PANI). The potential impact of the functional groups and porous shape of the electrospun nanofibrous electrode material on the enhancement of redox reactions in electrochemical measurements is worth considering. There exists the potential to advance the prototyping and miniaturization of the electrochemical configuration utilized in the examination of lung biomarker detection. This pilot study has demonstrated a new non-invasive and non-enzymatic approach for detecting lung cancer. To further expand on these findings, future research could involve conducting electrochemical measurements using standard volatile organic compounds (VOCs) such as decane, nonanal, and tetradecane. This would allow for the evaluation of the electrode's response to each individual VOC as well as their mixture. Additionally, the results obtained from these measurements could be correlated with real-world testing using Human Lung Adenocarcinoma A549 Cell lines as a model.

Acknowledgements We thankfully acknowledge DST-FIST, The Central Advanced Instrumentation Centre (CAIC), Government College for Women, Thiruvananthapuram, Kerala, Government project "Performance Linked Encouragement for Academic Studies" (PLEAS), Central Laboratory for Instrumentation and Facilitation (CLIF), Centre for Analytical Instrumentation—Kerala (CAI-K), KSCSTE-Kerala Forest Research Institute, Peechi, Thrissur, Kerala, Athmic Biotech Solutions Pvt. Ltd. Thiruvananthapuram, Kerala, and University of Kerala.

References

1. https://www.medicalnewstoday.com/articles/323648. Accessed the page on 17 April 2023
2. https://healthunits.com/cancer/what-is-lung-cancer. Accessed the page on 17 April 2014
3. Akitoshi S, Daiju U, Antoine C, Akira Y, Takashi H, Yuki S, Yukio M (2022) Deep learning-based algorithm for lung cancer detection on chest radiographs using the segmentation method. Sci Rep 12:727. https://doi.org/10.1038/s41598-021-04667-w4
4. Na Y, Jia Z, Fang C, Xiaokui T (2015) Circulating tumor cells in lung cancer: detection methods and clinical applications. Lung 193:157–171. https://doi.org/10.1007/s00408-015-9697-7
5. Lei H, Xiaocheng Y, Rongrong H, Lian J, Yuan L, Yan D, Li S, Hui C, Zhu C, Li Z, Pengfeng X, He N (2022) A novel specific and ultrasensitive method detecting extracellular vesicles secreted from lung cancer by padlock probe-based exponential rolling circle amplification. Nano Today 42:101334. ISSN 1748-0132. https://doi.org/10.1016/j.nantod.2021.101334
6. Shubhpreet K, Gagan J (2011) Watershed segmentation of lung CT scan images for early diagnosis of cancer. Int J Electr Comput Eng 850–852. https://doi.org/10.7763/IJCEE.2011.V3.431

7. Diksha M Kannan R, Ruchita T (2019) Deep learning algorithm for classification and prediction of lung cancer using CT scan images. In: 5th international conference on computing, communication, control and automation (ICCUBEA), Pune, India, pp 1 –5. https://doi.org/10.1109/ICCUBEA47591.2019.9128479
8. Aaon M (2021) An early prediction of lung cancer using CT scan images. J Comput Nat Sci 1:2. https://doi.org/10.53759/181X/JCNS202101008
9. Feifei M, Zhenyu L, Runkun Z, Gongke Li (2019) Cataluminescence sensor for highly sensitive and selective detection of ISO-butanol. Talanta 194:910–918. ISSN 0039-9140. https://doi.org/10.1016/j.talanta.2018.11.016
10. Shen X, Shi G, Zhang Y, Weng S (2022) Wireless volatile organic compound detection for restricted internet of things environments based on cataluminescence sensors. Chemosensors 10(5):179. https://doi.org/10.3390/chemosensors10050179
11. Varnakavi N, Nohyun L (2021) A review on biosensors and recent development of nanostructured materials-enabled biosensors. Sensors (Basel) 21(4):1109. https://doi.org/10.3390/s21041109. PMID: 33562639; PMCID: PMC791513
12. Camilla MC, Esther R, Eduardo E, Luis S, Alessio Z (2022) A paving the way for a green transition in the design of sensors and biosensors for the detection of volatile organic compounds (VOCs). Biosensors (Basel) 12(2):51. https://doi.org/10.3390/bios12020051. PMID: 35200311; PMCID: PMC8869180
13. Thriumani R, Zakaria A, Jeffree A, Hasyim YZ, Helmy KM, Omar MI, Shakaff AY, Kamarudin LM (2016) A study on VOCs released by lung cancer cell line using GCMS-SPME. Proc Chem 20:1–7. ISSN 1876-6196. https://doi.org/10.1016/j.proche.2016.07.027
14. Ratha S, Samantara AK (2018) Supercapacitor: instrumentation, measurement and performance evaluation techniques. Springer Briefs in materials. Springer, Singapore, vol 1, pp 1–47. https://doi.org/10.1007/978-981-13-3086-5
15. Borjigin M, Eskridge C, Niamat R, Strouse B, Bialk P, Kmiec EB (2013) Electrospun fiber membranes enable proliferation of genetically modified cells. Int J Nanomed 8:855–864. https://doi.org/10.2147/ijn.s40117
16. Sumitha MS, Xavier TS (2023) Synthesis and electrochemical characterization of electrospun biocompatible poly (ε-caprolactone) and polyaniline nanofiber composite electrode materials for various biosensing applications. Mater Today: Proc 80(2):1297–1305. https://doi.org/10.1016/j.matpr.2023.01.060
17. Rezaul KKM, Tarek M, Sarkar SM, Ong HR, Abdullah H, Cheng CK, Rahman KMM (2020) Selective synthesis of methanol by photoelectrocatalytic reduction of CO_2 over PANI-$CuFe_2O_4$ hybrid catalyst. IOP Conf Ser: Mater Sci Eng 736:04. https://doi.org/10.1088/1757-899X/736/4/042020
18. Ajeel KI, Sattar KQ (2019) Synthesis and characteristics of polyaniline (PANI) filled by graphene (PANI/GR) nano-films. J Phy: Conf Ser 1234:012020. https://doi.org/10.1088/1742-6596/1234/1/012020
19. Kartick LB, Krishna D, Arun B, Ranendu KN, Biswajit S (2016) Charge transport through polyaniline incorporated electrically conducting functional paper. J Phys Chem C 120:5855–5860. https://doi.org/10.1021/acs.jpcc.5b08650
20. Sagar HP, Aarti PG, Shivaram DS, Kashinath RP (2018) To form layer by layer composite film in view of its application as super capacitor electrode by exploiting the techniques of thin films formation just around the corner. Electrochim Acta 265:556–568. https://doi.org/10.1016/j.electacta.2018.01.165
21. Bogdan B, Andreea G, Paul D, Balan A, Valentin B (2017) Morphological and structural analysis of polyaniline and poly(o-anisidin) layers centered in a DC glow discharge plasma by using an oblique angle eectrode deposition configuration. Polymers 9:732–749. https://doi.org/10.3390/polym9120732
22. Leyla S, Mohammad K, Ali AE (2015) Preparation of electroactive nanofibers of star-shaped polycaprolactone/polyaniline blends. Colloid Polym Sci 293:481–491. https://doi.org/10.1007/s00396-014-3430-6

23. Abdelhaq B, Khalil J, Sylvain R, Claude D (2013) Grafting of polycaprolactone on oxidized nanocelluloses by click chemistry. J Nanomater 3(1):141–157. https://doi.org/10.3390/nano3010141
24. Sudha RP, Rajagopal K (2014) FTIR and Raman vibrational investigations on the complex of pyridine with tartaric acid. IOSR J Appl Phys (IOSR-JAP) e-ISSN: 2278-4861. 6, 4 Ver. III, 09–12. www.iosrjournals
25. Fábio FFG, Caitlyn AC, Pauline RH, Ranodhi WU, João CS, Paiyz EM, Carlos AVR, Joaquim MSC, Jorge MFM, Frederico CF, Robert JL (2019) Polyaniline-polycaprolactone blended nanofibers for neural cell culture. Eur Polym J 117:28–37. https://doi.org/10.1016/j.eurpolymj.2019.04.048
26. Mykola VS, Viktoriya SK, Viktor VS, Andrii SN, Genadiy PO, Volodymyr VV, Viktor VS (2014) Possibility of grapheme growth by close space sublimation. Nanoscale Res Lett 9:182–187. www.nanoscalereslett.com/content/9/1/182
27. Somruthai T, Kewalin I, Chanokporn C, Jantrawan P, Weerah W, Dheerawan B (2013) Improving the attachment and proliferation of umbilical cord mesenchymal stem cells on modified polystyrene by nitrogen-containing plasma. Cytotechnology 65(1):119–134. https://doi.org/10.1007/s10616-012-9467-9
28. Wagner CD, Naumkin AV, Kraut-Vass A, Allison JW, Powell CJ, Rumble JR Jr (2003) NIST standard reference database 20, Version 3.4 (web version) http://www.srdata.nist.gov/xps/. Cross Ref http://www.xpsfitting.com/2015/05/n-1s-for-organic-nitrogen-species.html. Accessed on 21 October 2021
29. Beamson G, David RB (1992) High resolution XPS of organic polymers: the Scienta ESCA300 database. Wiley Interscience
30. Jin CH, Lin YH, Paul WW (2020) X-ray photoelectron spectroscopy analysis of nitrogen-doped TiO_2 films prepared by reactive-ion-beam sputtering with various NH_3/O_2 gas mixture ratios. Coatings 10:47–60. https://doi.org/10.3390/coatings10010047
31. Kangjun X, Manman Z, Yang Y, Long Z, Wei Q (2018) Synthesis and supercapacitor performance of polyaniline/nitrogen-doped ordered mesoporous carbon composites. Nanoscale Res Lett 13:163–170. https://doi.org/10.1186/s11671-018-2577-3
32. Osadchii DY, Suarez AIO, Bavykina AV, Gascon J (2017) Revisiting nitrogen species in covalent triazine frameworks. Langmuir 33(50):14278–14285. https://doi.org/10.1021/acs.langmuir.7b02929
33. Jiajia X, Tong W, Yunqian D, Younan X (2019) Electrospinning and electrospun nanofibers: methods, materials, and applications. Chem Rev 119(8):5298–5415. https://doi.org/10.1021/acs.chemrev.8b00593
34. Md Shariful I, Bee CA, Andri A, Amalina M (2019) A review on fabrication of nanofibers via electrospinning and their application. SN Appl Sci 1:1248. https://doi.org/10.1007/s42452-019-12898-4
35. Chen X, Xu F, Wang Y, Pan Y, Lu D, Wang P, Ying K, Chen E, Zhang WA (2007) A study of the volatile organic compounds exhaled by lung cancer cells in vitro for breath diagnosis. Cancer 110(4):835–844. https://doi.org/10.1002/cncr.22844
36. Thriumani R, Jeffreea AI, Zakaria A, Hasyim YZHY, Helmy KM, Omar MI, Adom AH, Shakaff AY, Kamarudin LM (2015) A preliminary study on detection of lung cancer cells based on volatile organic compounds sensing using electronic nose. Jurnal Teknologi 77(7):67–71. https://doi.org/10.11113/jt.v77.6250
37. Filipiak W, Sponring A, Mikoviny T, Ager C, Schubert J, Miekisch W, Amann A, Troppmair J (2008) Release of volatile organic compounds (VOCs) from the lung cancer cell line CALU-1 in vitro. Cancer cell Int 8:17. https://doi.org/10.1186/1475-2867-8-17
38. Diana P, Paolo C, Massimo C, Matteo G, Olga A, Bruno B, Luca B, Michele R, Antonio M (2005) Exhaled volatile organic compounds in patients with non-small cell cancer, cross sectional and nested short term follow up study. Respir Res 6:71. https://doi.org/10.1186/1465-9921-6-71
39. Byun H, Yu J, Huh J, Lim J (2014) Exhaled breath analysis system based on electronic nose techniques applicable to lung diseases. Hanyang Med Rev 34:125–129. https://doi.org/10.7599/HMR.2014.34.3.125

40. Jin Y, Di W, Le W, Ping W, Yanjie H, Kejing Y (2009) Detection of lung cancer with volatile organic biomarkers in exhaled breath and lung cancer cells. AIP Conf Proc 23, 1137(1):198–201. https://doi.org/10.1063/1.3156506
41. Barash O, Peled N, Tisch U, Bunn PA, Hirsch FR, Haick H (2012) Classification of lung cancer histology by gold nanoparticle sensors. Nanomed Nanotechnol Biol Med 8(5):580–589. https://doi.org/10.1016/j.nano.2011.10.001
42. Patricia F, Christian L, Jochen KS, Wolfram M (2010) Breath gas aldehydes as biomarkers of lung cancer. Int J Cancer 126:2663–2670. https://doi.org/10.1002/ijc.24970
43. Mochalski P, Theurl M, Sponring A, Unterkofler K, Kirchmair R, Amann A (2014) Analysis of volatile organic compounds liberated and metabolised by human umbilical vein endothelial cells (HUVEC) in vitro. Cell Biochem Biophyd 71:323–329. https://doi.org/10.1007/s12013-014-0201-4
44. Changsong W, Ran D, Xiaoyang W, Ailing L, Chunjie C, Chaofu K, Lei G, Shanshan L, Wei Z, Guowang X, Enyou L (2014) Exhaled volatile organic compounds as lung cancer biomarkers during one-lung ventilation. Sci Rep 4(7312):1–8. https://doi.org/10.1038/srep07312

Study on Electrochemical Deposition of Thick Copper Pillars for Wafer Level Integration

Megha Agrawal, S. Harsha, Anila Puthoor, and S. Mahalakshmi

1 Introduction

Micro-Electro-Mechanical-Systems (MEMS) technology plays a significant role in microsystems as actuators and sensors. MEMS Packaging refers to providing encapsulation and providing electrical contacts to the die. If general MEMS and suitable Complementary Metal Oxide Semiconductor (CMOS) Integrated Circuit are integrated side by side, they consume more area. Hence for the miniaturization of a complete Microsystem, there is a push in industries for enhancing the packaging density. It can be achieved using advanced packaging techniques such as wafer-level 3-D integration and flip-chip bonding [1]. These techniques are also advantageous in terms of cost, yield, and reliability. It can be achieved by replacing single-chip wire bonding with metal bump bonding applications as the final step in chip manufacturing. Wafer-level integration includes wafer-to-wafer and chip-to-wafer stacking using Copper (Cu) pillars/bumps [2].

Among all the acceptable materials for interconnects, copper is the material of choice due to its excellent electrical conductivity, thermal conductivity, mechanical strength, CMOS process compatibility, and impressive electromigration resistance. Wafer-to-wafer integration can be performed using a wafer bonder; however, flip-chip bonding is helpful for chip-to-wafer integration and chip-to-package integration. Flip chip bonding reduces the cost of the device and improves its performance. There is a need for a higher pin count in a single micro-system, which demands reducing the pitch between pillars to realize the compact package. This can be achieved by replacing wire bonding with flip-chip bonding. These copper pillars can be formed

M. Agrawal (✉) · S. Harsha · A. Puthoor · S. Mahalakshmi
Sensor Technology Development Centre, Centre for Sensors & Vision Technology, Central Manufacturing Technology Institute, Bangalore, India
e-mail: megha@cmti.res.in

S. T. Aruna et al. (eds.), *Recent Advances in Electrochemical Science and Technology*, Springer Proceedings in Materials 47, https://doi.org/10.1007/978-981-97-3958-5_10

Fig. 1 Electroplating setup

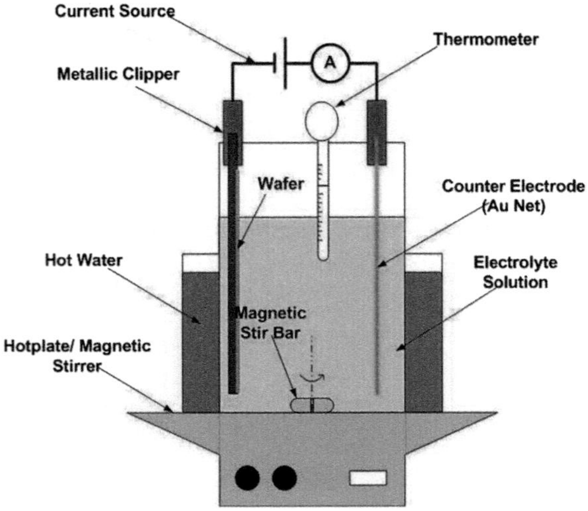

at the wafer level using photolithography and electroplating processes. The electroplating process, a well-stabilized technology for micro-features, is used to create these solder bump/copper pillars using a photoresist mold [3].

Electroplating has been a well-established technology for industrial and commercial applications since the nineteenth century to deposit thick metal layers on the conductive surface. Up to 100 microns thick copper can be deposited using this technique with good quality and acceptable internal stresses [4].

Typically an electroplating system cell consists of two electrodes, i.e., anode and cathode, which are usually made of metal and held apart in the electrolyte (Fig. 1). The sample (to be plated) with a conductive surface is connected to the cathode of a power supply, while an electrode made of the target material is connected to the anode. Both electrodes and wafers are then immersed in the electrolyte solution, which contains ions of copper. A constant concentration of Cu^{2+} is maintained in the electrolyte solution surrounding the electrodes:

$$\text{Anode: } Cu(s) \rightarrow Cu^{2+}(aq) + 2e^{-}$$

$$\text{Cathode: } Cu^{2+}(aq) + 2e^{-} \rightarrow Cu(s)$$

Photolithography (Fig. 2) and electroplating can be used to grow copper pillars selectively. Selective electroplating is advantageous as it avoids the etching step to remove the unwanted metal and uses an electroplating solution optimally. Various photoresists are used based on the thickness requirement for metal (to be electroplated). Since copper balls are used for interconnects, it is required to create tall and thin copper pillar structures with smaller pitches to enhance the packaging density. The tall copper pillar structures eliminate the requirements of underfilling. With thin photoresists like AZ1518 and AZ5214, a mold height of only up to 2 μm with a

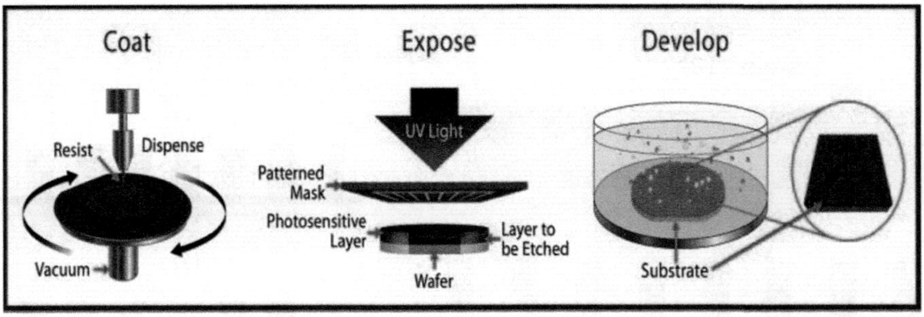

Fig. 2 Photolithography process steps

single coat can be achieved. These results are mushroom or umbrella-type over-plating bumps, which cannot be used for high bump count devices. There is a need to create a mold with a thick photoresist with a typical thickness greater than 50 μm [5, 6]. To overcome this issue, thicker film may be achieved using multiple coating processes of a thin photoresist, longer soft bake times, and higher exposure dose to achieve the required thickness with a high aspect ratio, which increases the processing time and hence the process cost. The coating and pattering of these thick photoresist films with high aspect ratio structures are highly challenging.

The process of these thick photoresists with fine pitch is significantly different and challenging from thin photoresists such as AZ1518 and AZ5240, which are used to fabricate MEMS devices [4–6]. SU-8 photoresist fulfills the demand of fabricating thicker structures in a single coating step with a high aspect ratio [7–9]. However, since the SU-8 will be highly cross-linked after hard baking, it is difficult to strip off the SU-8 using a standard stripping solution after curing. This process issue limits the use of SU-8 in the application of selective electroplating, where the photoresist is used as a mold material and has to be removed after the process is necessary. Hence a thick photoresist with similar performance to SU-8 but enhanced strip ability using standard wet photoresist strip chemistry would be highly desirable. AZ10XT and AZ125 are thicker acrylic photoresists and can be stripped off easily after the electroplating process [10].

The main objective of this work is to create photoresist molds of different thicknesses for the formation of copper pillars with electroplating of pitch and diameter down to 15 μm. This proposed approach is also helpful for various applications like high aspect ratio on-chip inductors, micro-gear, cantilevers, valves, etc.

2 Experimental

The steps involved in the selective electroplating process are depicted in Fig. 3. A 6-inch P-type <100> silicon wafer was first cleaned with piranha solution for 10 min at 120 °C. The piranha solution is a mixture of concentrated sulfuric acid (H_2SO_4) and

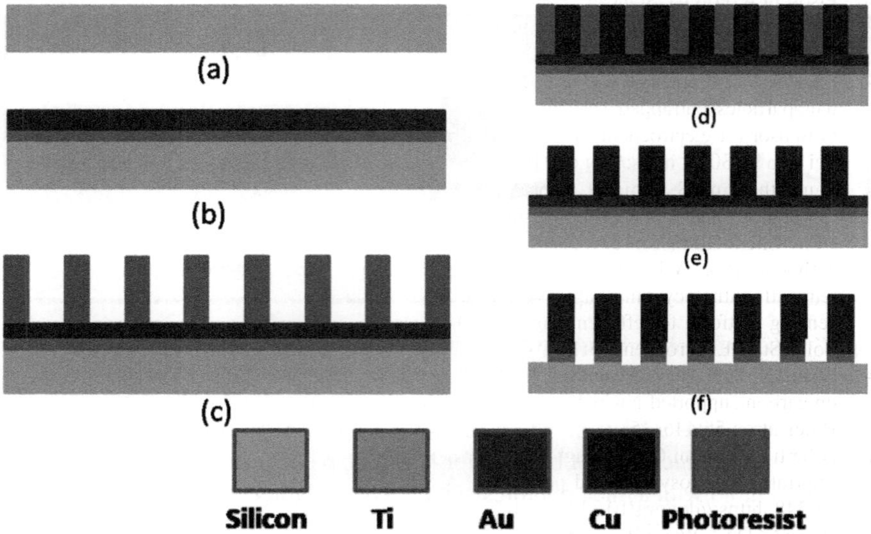

Fig. 3 Selective electroplating process flow : (**a**) Silicon wafer cleaning, (**b**) Si wafer coated with Ti/Au seed layer, (**c**) photoresist spin coating followed by lithography, (**d**) copper electroplating, (**e**) photoresist striping, (**f**) patterning and etching of Ti/Au seed layer

hydrogen peroxide (H_2O_2) in a ratio of 3:1 to remove organic contaminants from the surface. After cleaning, a thin layer of titanium (Ti) with 20 nm thickness was coated to increase the adhesion of the gold layer with silicon. Above this, a gold (Au) layer of 50 nm was deposited. Both Ti and Au layers were deposited through an electron beam evaporation system. The Au acts as a seed layer for copper electroplating to ensure the adhesion of copper to the substrate as well as to increase the copper growth stability during the process. These 6-inch Silicon wafers with 20 nm titanium and 50 nm gold were diced into 25 mm × 25 mm pieces. These diced pieces were used as substrates for all the copper pillar experimental work reported here.

To selectively deposit the copper metal, thick photoresist molds must first be created with a photolithography process. Here all lithography parameters are optimized for thick positive photoresist AZ10XT and negative photoresist AZ125nXT. General cleaning was done on the Au-coated silicon sample using acetone, isopropyl alcohol, and deionized water to remove the contaminants and dried using a nitrogen gun, then dehydrated at 200 °C for 10 min using a hot plate. After cleaning, the AZ10XT photoresist was spin-coated on a silicon sample with a spin speed of 2400 rpm for 60 s and soft-baked for 165 s at 110 °C (Table 1).

The direct writing of circular patterns was done using a laser writer (MicroTech) with an exposure dose of 1130 mJ/cm^2 and 1309 mJ/cm^2 for single and double coatings, respectively. These are then developed with AZ726MIF development solution and hard baked for 5 min at 110 °C. Similarly, AZ125nXT was coated on Au-coated silicon sample with a spin speed of 1600 rpm for 10 s. Since AZ125nXT is thicker than AZ10XT, it requires more soft baking and more exposure dose. Soft baking is

Table 1 Photolithography parameters for AZ10XT and 125nXT

Sample	S1	S2	S3
Photoresist	AZ10XT	AZ10XT	AZ125nXT
No. of spin cycles	Single	Double	Single
Spin speed (rpm)	2400	2400	1600
Spin time (s)	60	60	10
Thickness targeted (μm)	8	16	50
Exposure dose (mJ/cm^2)	1130	1309	1635

done for 20 min at 130 °C and exposed to 1635 mJ/ cm^2 dose. All photolithography parameters are listed in Table 1. Features written with laser writer were 20, 30, 40, and 50 μm in diameter and 20, 30, 40, and 50 μm gap between the pillars. The design parameters are mentioned in Table 2.

The copper electroplating was carried out in Mu-IKO station from ECSI Fibro tools Inc. USA using horizontal cathode and anode holder. The acidic copper electrolyte of high aspect ratio, prepared in the laboratory, contains three primary inorganic components and two organic compounds. The inorganic components: copper sulfate ($CuSO_4$) provides a source of copper ions. Sulfuric acid (H_2SO_4) makes the bath conductive and acts as a charge carrier. Hydrochloric acid (HCl) increases the solution conductivity and as well as it prevents anode passivation. These chloride ions (Cl^-) combine with the carrier (organic compound) and form a complex which controls the plating rate in selective area. The brightener solution is a grain refiner that allows uniform, low-stress, fine-grained deposition. The composition of copper electrolytes is given in Table 3.

There are three different ways of plating copper films, viz., direct current method, forward pulse plating, and reverse pulse plating (Fig. 4). In DC plating, the current is continuous in copper, resulting in columnar grains with increased porosity and hardness. It is also unsuitable to fill deep gaps due to its poor throwing power capacity. The advantage of pulse plating is that copper ions will be replenished near the cathode

Table 2 Design parameters

Sl. no	Diameter of pillars (μm)	Gap between the pillars (μm)	Array	Area of a single pillar (10^{-12}) m^2	Total no. of pillars
S1	10	10	500 × 500	314	250,000
S2	20	20	400 × 400	1256	160,000
S3	30	30	300 × 300	2826	90,000
S4	40	40	200 × 200	5024	40,000
S5	50	50	100 × 100	7850	10,000

Table 3 Composition of one liter copper electrolyte

Chemicals required	Quantity (mL)
Purified liquid copper sulfate ($CuSO_4 \cdot 5H_2O$)	278
Concentrated sulfuric acid (H_2SO_4)	120
Concentrated hydrochloric acid	0.13
Brightener	7.0
Carrier	7.0
Deionized water	588

surface during the off-cycle time, making coatings more uniform, with better hardness and low porosity than DC plating. Reverse pulse plating can dissolve back copper in reverse on a time cycle, making coatings uniform, especially in the case of complex structures like TSV and copper pillars.

In this work, copper pillars are deposited by the pulse reverse plating method. An acidic copper electrolyte bath was used for the electrochemical deposition, with a soluble anode to maintain the copper ions concentration in the bath. According to Faraday law, the deposition amount is proportional to the current and time applied. Here, a 20 μm/h deposition rate is achieved with a forward current density of 20 mA/cm^2 and a reverse current density of 60 mA/cm^2 (Table 4). After the copper deposition, photoresist stripping was done using hot acetone for 20 min.

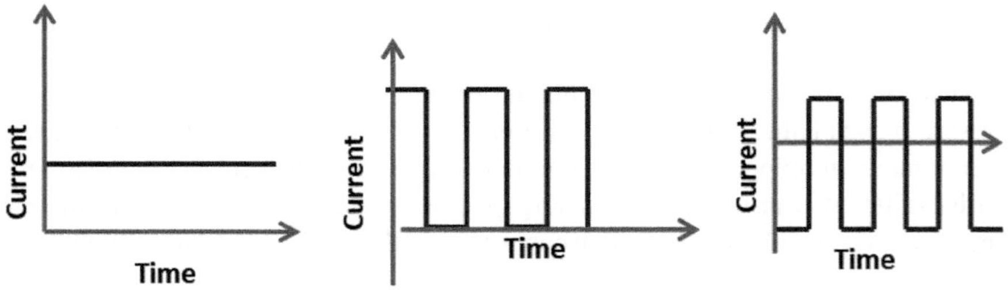

Fig. 4 Electroplating power waveforms **a** DC, **b** pulse forward, **c** pulse reverse

Table 4 Electroplating parameters

Parameters	Value
Forward current density	20 mA/cm^2
Reverse current density	60 mA/cm^2
Total area	0.7857 cm^2
Deposition rate	20 μm/h
pH of bath	2.6
Temperature of bath	30 °C
Anode acceleration speed	8 rpm

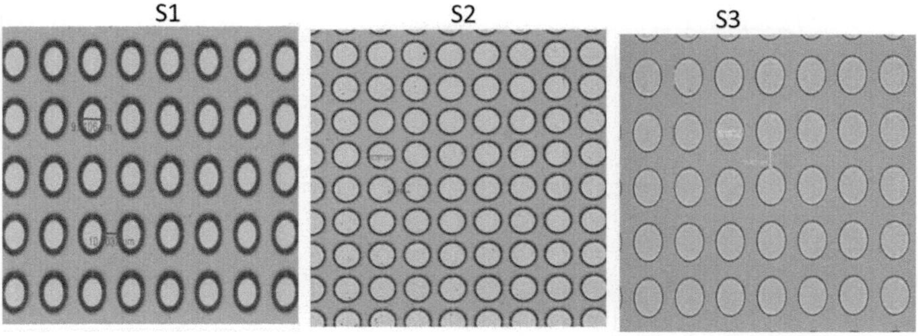

Fig. 5 Microscopic images after photolithography

3 Results and Discussions

After photolithography, the optical images of photoresist molds are captured (Fig. 5) using an optical microscope (Leica DM8000M), and thickness was measured using Form Talysurf (PGI Series 1500S). The thickness values of 9.3 μm and 14.8 μm were obtained with AZ10XT for the single spin coating cycle and double coating cycle respectively (Fig. 6), while with AZ125nXT, it is 48.9 μm.

Figure 7 shows the optical images of the copper pillars and they are uniformly seen. This has been also substantiated by the confocal microscope images of the copper pillars (Fig. 8) and FESEM images (Fig. 9).

4 Conclusions

The methodology for the practical implementation of copper pillars through selective electroplating was achieved. This methodology was successfully established by considering several influencing parameters of the processes, viz., lithography, electrodeposition, etc., and by optimizing each of those parameters. Targeted development of copper pillars with a minimum diameter of 10 μm and other diameters of 20 and 30 μm were also achieved successfully. Process compatibility with both thin and thick photoresists was verified and implemented by modifying the process parameters suitably. Preprocesses for wafer-level integration of copper pillars were established. The results indicated the successful formation of copper pillars.

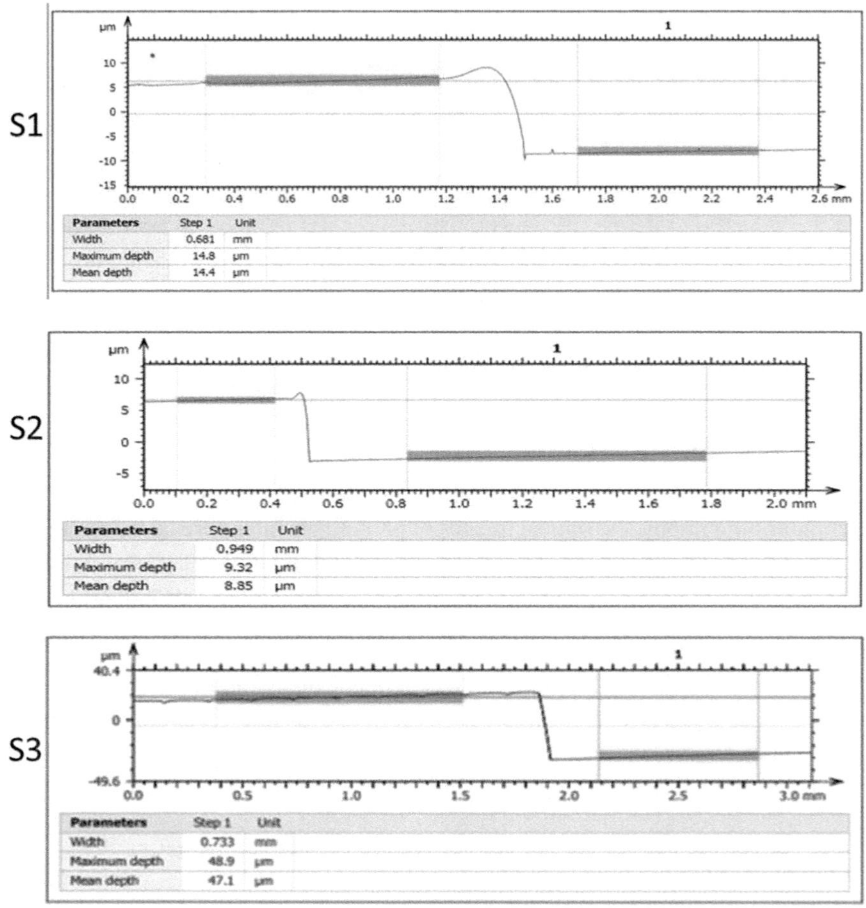

Fig. 6 Profile measurement of photoresists molds created using photolithography

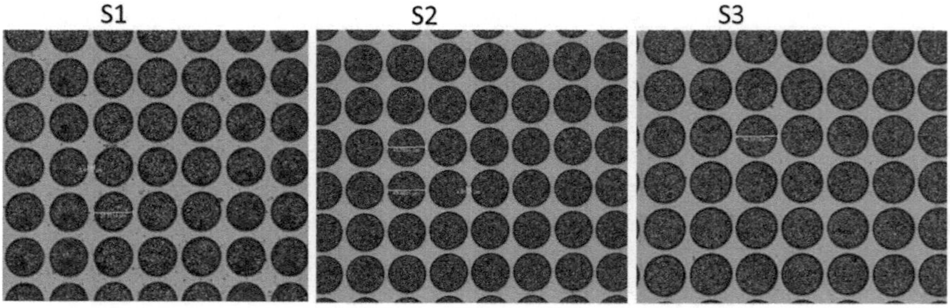

Fig. 7 Optical images of copper pillars

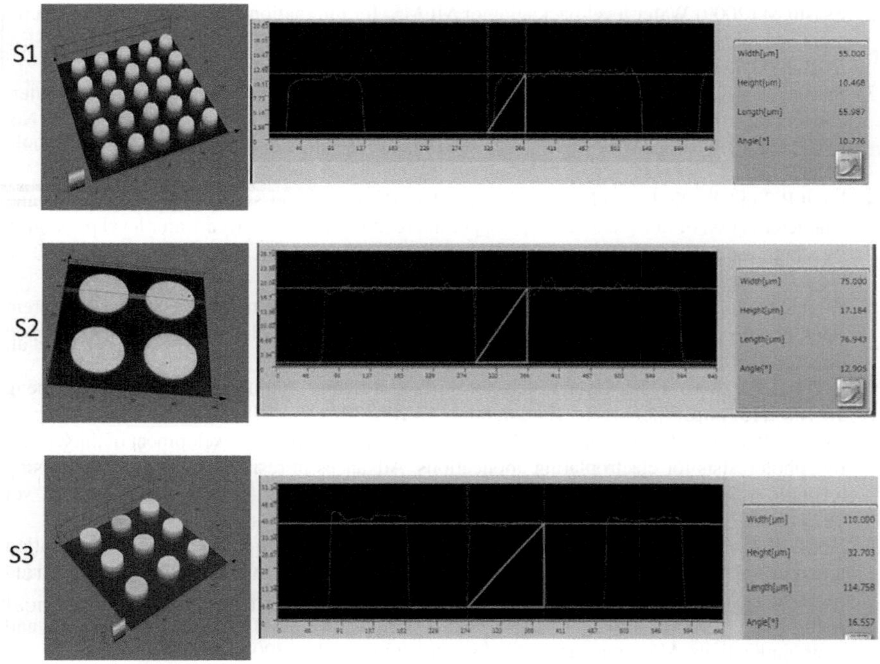

Fig. 8 Confocal images of electroplated copper pillars

Fig. 9 SEM images of the electroplated copper pillars

Acknowledgements The authors would like to thank the Sensor Technology Development Centre staff and other labs of CMTI for their cooperation and help in the experimental work. This work is funded by the Department of Heavy Industries (DHI) through Sensor Technology Development Facility (STDF) project.

References

1. Lu J-Q (2009) 3-D hyper integration and packaging technologies for micro-nano systems. Proc IEEE 97(1)

2. Esashi M (2009) Wafer level packaging of MEMS. In: International solid-state sensors, actuators and microsystems conference, Denver, CO, USA, pp 9–16. https://doi.org/10.1109/SENSOR.2009.5285574
3. Shimokawa K, Hashino E, Ohzeki Y, Tatsumi K (1998) Micro-ball bump for flip chip interconnections. In: Proceedings 48th electronic components and technology conference (Cat. No. 98CH36206), Seattle, WA, USA, 1998, pp 1472–1476. https://doi.org/10.1109/ECTC.1998.678939
4. Dixit P, Tan CW, Xu L, Lin N, Miao J, Pang JHL, Backus P, Preisser R (2007) Fabrication and characterization of fine pitch on-chip copper interconnects for advanced wafer level packaging by a high aspect ratio through AZ9260 resist electroplating. J Micromech Microeng 17:1078. https://doi.org/10.1088/0960-1317/17/5/030
5. Koukharenko E, Kraft M, Ensell GJ, Hollinshead N (2005) A comparative study of different thick photoresists for MEMS applications. J Mater Sci: Mater Electron 16:741–747. https://doi.org/10.1007/s10854-005-4977-2
6. Loechel B (2000) Thick-layer resists for surface micromachining. J Micromech Microeng 10:108–115. https://doi.org/10.1088/0960-1317/10/2/302
7. Chen C, Plass B, Edward NG, Lee S, Meyer S, Pawlowski G (2008) Development of thick negative photoresists for electroplating applications. Advances in resist materials and processing technology. In: Proceedings of SPIE - the international society for optical engineering, vol 6923:69233E. https://doi.org/10.1117/12.773986
8. Hoa C-H, China K-P, Yangb C-R, Wuc H-M, Chenc S-L (2002) Ultrathick SU-8 mold formation and removal, and its application to the fabrication of LIGA-like micromotors with embedded roots. Sens Actuators A 102:130–138. https://doi.org/10.1016/S0924-4247(02)00297-2
9. Jeong Bong L, Kyung-Hak C, Koangki Y (2015) Innovative SU-8 lithography techniques and their applications. Micromachines 6:1–18. https://doi.org/10.3390/mi6010001
10. Sampada Sameer N, Harsha S, Megha A, Mahalakshmi S (2020) Electrochemical deposition and characterization of copper pillar bumps—application towards flip chip bonding. In: Proceeding of 5th IEEE international conference on emerging electronics (ICEE). IEEE. https://doi.org/10.1109/ICEE50728.2020.9777005

Effect on Morphology and Corrosion Behavior of AlCrFeCoNi High-Entropy Alloy Coatings Prepared by Electroless Process on Mild Steel Substrate

Mansha Mohan, Sujata Singh, Upender Pandel, and Krishna Kumar

1 Introduction

In 2004, Yeh [1] proposed a new class of alloys, namely, high entropy alloys (HEAs) consisting of five or more principal elements along with one primary element, unusual to conventional alloys. For near equiatomic HEA, high mixing entropy, ΔS_{mix} can be expressed as $\Delta S_{\text{mix}} = R \ln n$, ($R$ is gas constant and n is total no of constituent elements) which stabilizes the formation of solid solution phase rather than intermetallic compounds [2]. Four core effects in HEAs such as the high entropy effect, sluggish diffusion, lattice distortion, and cocktail effect will alter the morphological and compositional properties. HEAs coatings are very effective in the field of surface engineering in studying corrosion behavior, chemical oxide stability, and texture performance for improvizing desirable properties such as tribological performance, corrosion behavior, and hardness [3]. Among various coating processes, electroless deposition is simple and cost-effective which is widely used and is now widely acknowledged by many industries to enhance several properties. Incorporation of several defects such as pores, local segregation, and oxidation can be observed in electroless deposited coatings. Although many types of heat treatments are done for improvized properties [4], however, post annealing can be an effective route to study for improvizing several surface properties of the coatings. HEA coatings through plasma spray route, laser cladding, and detonation spraying technique, have been studied after post annealing treatment to improve microstructure evolution, wear resistance, and phase stability. Wang et al. [5] studied AlCrCoFeNi HEA coating by plasma spray process for microstructural evolution and mechanical behavior. The

M. Mohan (✉) · U. Pandel · K. Kumar
Metallurgical and Materials Engineering, Malaviya National Institute of Technology, Jaipur, Rajasthan 302017, India
e-mail: 2019rmt9040@mnit.ac.in

S. Singh
Department of Materials Engineering, Indian Institute of Science, Bangalore 560012, India

S. T. Aruna et al. (eds.), *Recent Advances in Electrochemical Science and Technology*, Springer Proceedings in Materials 47, https://doi.org/10.1007/978-981-97-3958-5_11

lamellar microstructure was observed in as coated samples; however, no further changes were observed after annealing but the hardness increased from 468 to 575 HV. The presence of BCC structure might possibly enhance the hardness value [6]. Certain advances in results inspired to carrying out further studies related to AlCr-CoFeNi HEA coatings and enhance corrosion-resistant properties of mild steel. Many HEA coatings have been developed through the electrodeposition route by incorporating several carbonaceous elements like graphene oxide and carbon nanotubes to study the corrosion behavior of the coatings but electroless deposition is a very rare field of research in HEAs coatings on mild steel [7, 8]. Hence, this study reports the development of AlCrCoFeNi HEA coating by electroless process and investigation of its morphology, phase stability, and corrosion behavior at high temperature, before and after post annealing process on a mild steel substrate.

2 Experimental Procedure

2.1 Electroless Deposition

A mild steel sample (20 mm × 20 mm × 0.9 mm) was used as the substrate for the electroless deposition of AlCrFeCoNi HEA coatings and followed by annealing at 600 °C temperature for 30 min. The mild steel substrate, coated sample, and post annealed sample were designated as MS, HEA_1, and HEA_2, respectively. The mild steel substrate was ground and polished using 80–1200 grit size emery papers and ultrasonicated in acetone for 30 min. The samples were activated using dil. HCl for 20 is followed by rinsing with DI water before deposition. All the metal salts were used in chloride forms for preparing electrolytic bath along with additives. Sodium hypophosphite monohydrate (2.5 g/L) was used as the reducing agent and lactic acid (20 mL/L) as a complexing agent in the bath. HCl solution was used to maintain the electrolytic bath at 1.5 pH. The electroless deposition was carried out using bath composition and operating conditions as tabulated in Table 1. After electroless coating, heat treatment was carried out in an STF1200 tube furnace of thermal CVD operated with a flow rate of 200 cm and a pressure of 12 Torr. Before annealing, purging was done 3 times for 5 min each in argon environment. Annealing was carried out in an inert atmosphere at a heating rate of 7–8 °C/min, holding for 30 min at 600 °C, and the cooling rate was kept at 3–4 °C/min.

2.2 Characterization and Electrochemical Analysis

The composition and coating morphology were carried out using Nova Nano SEM 450, FEI Scanning Electron Microscope (SEM) equipped with energy dispersive X-ray spectrometer (EDS). Surface roughness was calculated using the Profilm model,

Table 1 Electroless deposition bath conditions used for coatings deposition

Sample	Bath composition	Concentration (g L^{-1})	Bath condition
HEA_1 (AlCrFeCoNi HEA coating)	$AlCl_3 \cdot 6H_2O$	24.14	Temperature: 80 °C Deposition time: 45 min pH: 1.5, under continuous stirring
	$CoCl_2 \cdot 6H_2O$	9.52	
	$CrCl_3 \cdot 4H_2O$	33.30	
	$FeCl_2 \cdot 4H_2O$	6.96	
	$NiCl_2 \cdot 6H_2O$	19.02	
HEA_2	Annealing at 600 °C		

a 3D profilometer apparatus. Vertical scan and back scan length were 200 μm and 20 μm, respectively. To calculate the coating thickness, cross-section was polished using an Illion II instrument and micrographs were reported by SEM. Phase identification was performed through a Panalytical XPert Pro instrument with a variation of 2θ from 20 to 120° with *Cukα* ($\lambda = 1.54$ Å) as the source X-ray. Water contact angle measurements were carried out by contour analysis system using OCA 15EC model. All contact angle measurements were done after the water droplet was stable for 20 s on the coated surface. An average value of three different measurements was reported.

Electrochemical studies were performed in 3.5% NaCl solution for the as-deposited coatings using PG-Lyte 1.0 potentiostat/galvanostat. A three-electrode system consisting of coated mild steel as the working electrode, platinum mesh as the counter electrode, and saturated calomel electrode (SCE) as the reference electrode. Electrochemical impedance spectroscopy (EIS) was executed using 0.005 V AC voltage amplitude after exposing the coated area of 1 cm^2 for one hour to stabilize the OCP value (open circuit potential). EIS was carried between the frequency range of 10^{-2}–10^5 Hz. The analysis of the EIS data and curve fitting of measured data was done by Zsimp 3.2.1 software and the electrochemical data was performed at room temperature at least four times for reproducibility of the results. The XPS spectra were carried out using the Omicron Nano Technology instrument. The analysis of XPS spectra for all the elements was done by taking carbon as the reference and employing CASA software.

3 Results and Discussion

3.1 Morphology, Composition, and Phase Analysis

Surface morphological images of mild steel (MS), as coated (HEA_1), and post annealed samples (HEA_2) are shown in Fig. 1. The morphology of mild steel was taken after polishing the substrate up to 2000 grit size emery paper. The as-coated AlCrFeCoNi HEA shows a continuous and uniform surface while the annealed

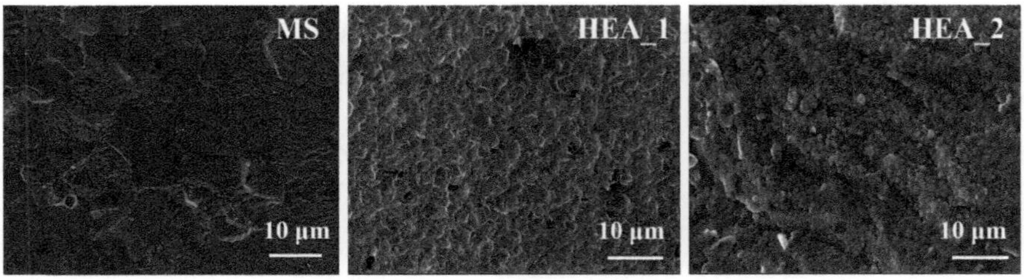

Fig. 1 SEM morphology of as coated AlCrFeCoNi (HEA_1) and post annealed (HEA_2) samples

sample shows denser and more granular morphology [9]. As the coating is annealed at 600 °C, the sharp edges in splats form became globular and smooth which indicates that the AlCrFeCoNi HEAs are recrystallized. Figure 2a shows the elemental at.% of all the AlCrFeCoNi elements present in the HEA_1 and HEA_2 samples through SEM–EDS analysis. Insignificant differences were observed in EDS before and after annealing. The variation of the elements is between 5 and 35 at.% which confirms the presence of all the elements and formation of high entropy alloy as tabulated in Table 2. The distribution of all the elements after electroless deposition can be observed through elemental mapping for HEA_1 as shown in Fig. 2b which also confirms the presence of elements. Figure 3 shows a top view and 3D view for HEA_1 and HEA_2 samples depicting their roughness. The observation shows that the surface roughness is decreased for HEA_2 than that for HEA_1. The roughness value gets reduced from 1.218 ± 0.32 μm for HEA_1 to 0.7112 ± 0.67 μm for HEA_2. Many researchers suggested that the corrosion resistance can be improved with a decrease in surface roughness value. The higher the roughness, the higher the surface area that can act as a nucleation site for the corrosive media to attack leading to more corrosion [10]. Hence, after heat treatment, fine-grained, smooth granules, and compact morphology are formed resulting in lower roughness of the film surface and subsequently improving the corrosion resistance. This can also be further related to the corrosion analysis later in Sects. 3.4 and 3.5.

Figure 4 shows the X-ray diffractograms of the substrate, HEA_1 and HEA_2. The substrate showed intense peaks corresponding to Fe with (h k l) values (1 1 0), (2 0 0), (2 1 1), (2 2 0), and (310) forming BCC phase structure at 2θ angles of 44.72°, 65.10°, 82.5°, 98.81°, and 116.35°, respectively. The peak intensities for HEA_1 and HEA_2 revealed Cr-Fe-Ni and AlNi compounds respectively which form BCC solid solution phase structure showing more sharp intensity peaks. This can also be observed through other experimental observations for AlCrFeCoNi HEA multicomponent systems because of its high configurational entropy [5]. The BCC phase fraction increased for the heat-treated coating at 600 °C and the diffraction peak intensity also increased. The change in phase fraction can be clearly seen in the deconvoluted peaks (2θ range: 81°–84°, 98°–100°, 115°–118°) for both HEA_1 and HEA_2 in Fig. 5. Hence, no phase transformation was observed after heat treatment (600 °C), which indicate the phase stability at high temperature [11]. Further, to investigate the microstructural evolution of coatings, before and after heat treatment,

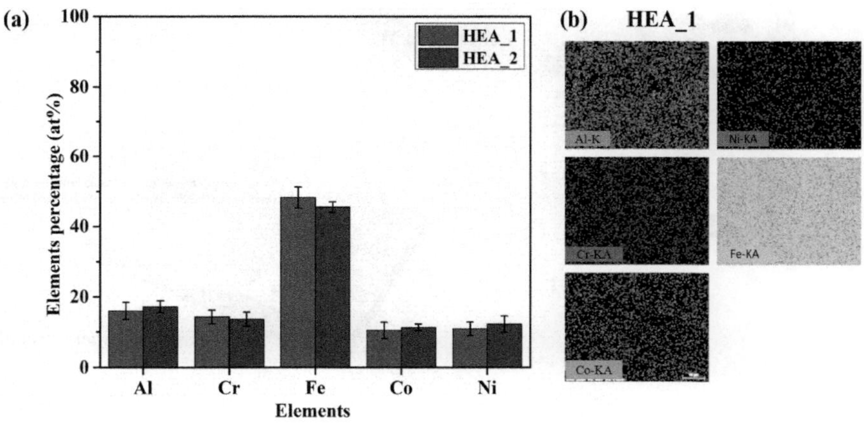

Fig. 2 **a** SEM–EDS elemental composition (atomic %) and **b** EDS mapping of element distribution for as deposited AlCrFeCoNi HEA coating

Table 2 Average composition of the elements present in AlCrFeCoNi, HEA_1, and HEA_2 system, analyzed by SEM-EDS

System/elements	Al	Cr	Fe	Co	Ni
HEA_1 (at.%)	16.02 ± 2.45	14.30 ± 1.94	48.26 ± 3.02	10.48 ± 2.34	10.94 ± 1.98
HEA_2 (at.%)	17.21 ± 1.71	13.67 ± 2.03	45.53 ± 1.52	11.34 ± 0.94	12.25 ± 2.34

the average crystallite size was calculated using the Scherrer equation [12] ($D = K\lambda/(\beta \cos\theta)$), where D is Crystallite size, K is shape factor (0.9), θ is diffraction angle, λ is the wavelength of *Cukα* radiation source. The lattice strain was calculated using the equation, $\varepsilon = \beta/4 \tan\theta$. The average crystallite size slightly increased from HEA_1 (33.92 nm) to HEA_2 (34.06 nm) while the observed lattice strain decreased from 2.09 to 1.66. The annealing treatment carried out can lead to crystallization and grain growth in HEAs [13] despite, high entropy effect can result in insignificant change in the crystallite size at 600 °C. The crystallinity of films improved as evident from the increased overall peak intensities of both Cr-Fe-Ni and AlNi present. The increase in crystallite size can also be correlated to the growth in crystallite size after heat treatment. Also, microstrain value decreases with increasing crystallite size as annealing is strain-relieving heat treatment. Since defects get eliminated with the increase in annealing temperature, microstrain in the film was reduced at 600 °C [13]. Hence, at 600 °C, the phase stability can be achieved with enhancement in corrosion resistance and is discussed later in Sects. 3.4 and 3.5.

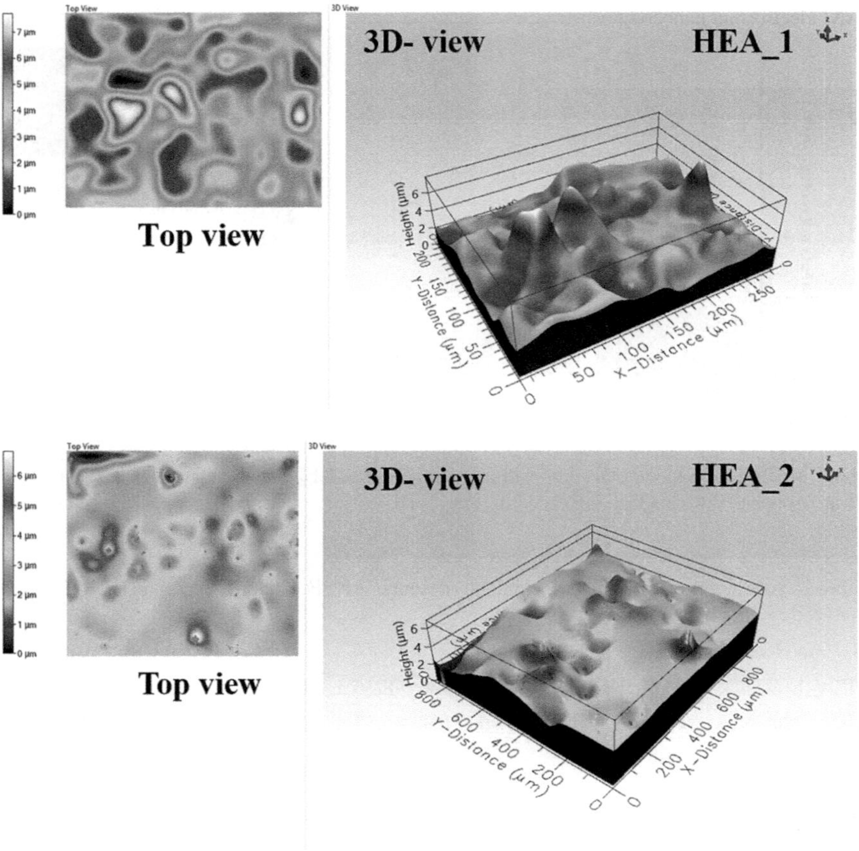

Fig. 3 2D and 3D surface profiles of HEA_1 and HEA_2

Fig. 4 X-ray diffractograms of MS, HEA_1, and HEA_2 samples

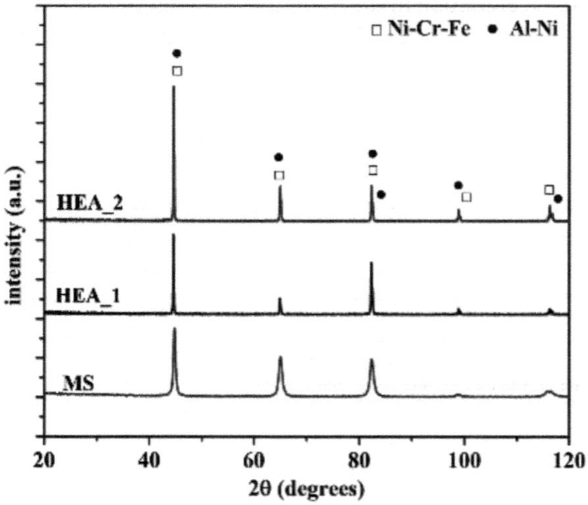

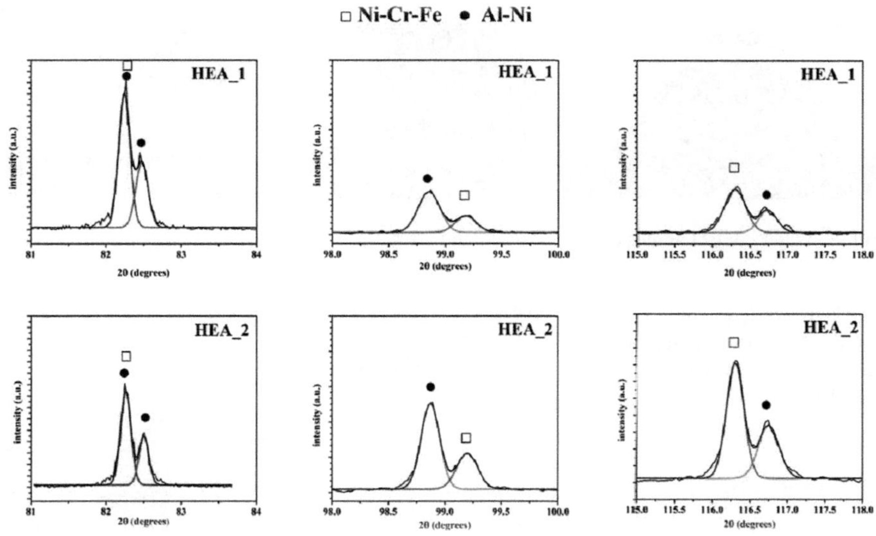

Fig. 5 Deconvolution of XRD peaks for HEA_1 and HEA_2 samples

3.2 *Wettability (Contact Angle Measurement)*

Water contact angles were measured to study the wettability of the surfaces of the mild steel substrate, HEA_1, and HEA_2. Figure 6 displays contact angles of $76.12 \pm 1.44°$, $125.26 \pm 1.08°$, and $141.35 \pm 0.98°$ for MS, HEA_1, and HEA_2, respectively. The water contact angle of MS was <90° showing the hydrophilic nature of the substrate but, the HEA coating shows a hydrophobic nature (>90°). Furthermore, annealing of the coating resulted in increased hydrophobicity to a greater extent than that of the substrate. Figure 7 shows the plot of the contact angle and fraction of air pockets which can be calculated using the Cassie-Baxter equation [14]: $\cos \theta r = fs \cos \theta - fa$, where θ: contact angle of the substrate, θr: contact angle of the coated samples, fs: fraction of solid surface, and fa: fraction of air at the interface ($fs + fa = 1$). The fraction of air pockets for HEA_1 was 65%, whereas for HEA_2, it was 82% (Fig. 7). The higher fraction of air pockets prevents the penetration of corrosive media throughout the coating forming a hydrophobic layer and thus preventing the corrosion. The wettability of the surface can be related to high entropy alloy formation and globular structure of the coatings. The enhanced corrosion property is due to the strong hydrophobic nature which has been further discussed in Sect. 3.7.

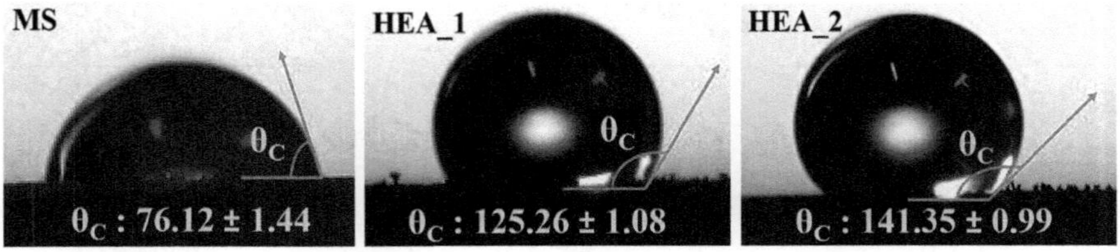

Fig. 6 Water contact angles of MS, HEA_1, and HEA_2

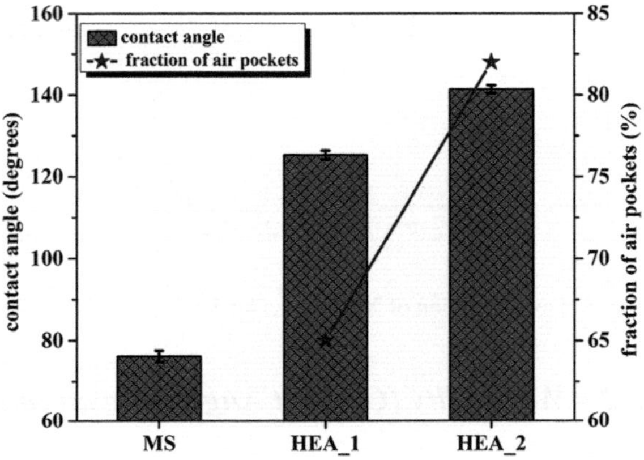

Fig. 7 Comparison of contact angle and fraction of air pockets of MS, HEA_1, and HEA_2

3.3 Criteria to Form High-Entropy Alloy

HEAs are most likely to form simple solid solution phases such as BCC, FCC, HCP, etc., with the most reassuring properties. Gibbs free energy of mixing of liquid phase (ΔG) is an important consideration for solid solution formation in HEA alloys which can be written as:

$$\Delta G_{\text{min}} = \Delta H_{\text{mix}} - T\Delta S_{\text{mix}} \tag{1}$$

Table 3 demonstrates the high entropy alloy parameters: δ, ΔH_{mix}, ΔS_{mix}, T_m, and Ω calculated for the AlCrFeCoNi HEA system which shows high entropy alloy formation according to previous studies [15]. Where δ: atomic size difference, ΔH_{mix}: enthalpy of mixing (kJ mol^{-1}), ΔS_{mix}: entropy of mixing (J K^{-1} mol^{-1}), T_m: absolute mixing temperature (K), Ω: parameter to predict solid solution formation, c_{ij}: atomic percentage of ith component $[\sum_{i=1}^{n} c_{ij} = 1]$, r_i: atomic radius, $\bar{r}$: average atomic radius, and R: gas constant (8.314 J K^{-1} mol^{-1}). The entropy of mixing should be maximum for equiatomic ratio alloy. Also, all the elements present in the system should be present between 5 and 35% at.% which has also been discussed earlier in section through SEM-EDS analysis. The phase formation is affected by many factors in multicomponent alloy systems. A larger binding force exists between the

Table 3 Calculated δ, ΔH_{mix}, ΔS_{mix}, T_m, Ω parameters for AlCrFeCoNi HEA system [15]

Parameters	Formula	Range of HEA solid solution	Reported values
δ	$\sqrt{\sum_{i=1}^{n} c_i\left(1-\frac{r_i}{r}\right)^2}$	$\delta < 6.5\%$	5.25%
ΔH_{mix}	$\sum_{i=1, i\neq j}^{n} \Omega_{ij} c_i c_j$	-15 kJ/mol < ΔH_{mix} < 5 kJ/mol	-12.32 kJ/mol
ΔS_{mix}	$-R\sum_{i=1}^{n} c_i ln c_i$	$\Delta S_{\text{mix}} > 1.5\text{R}$	13.38 J K^{-1} mol^{-1}
T_m	$\sum_{i=1}^{n} c_i (T_m)_i$	–	1675.10 K
Ω	$\frac{T_m \Delta S_{\text{mix}}}{\lvert\Delta H_{\text{mix}}\rvert}$	$\Omega > 1$	1.83

elements when ΔH_{mix} exhibits more negative value. The ΔH_{mix} for the studied HEA system is -12.32 kJ/mol which demonstrates random element distribution forming a solid-solution phase. The high mixing of entropy (ΔS_{mix}) also accounts for the solid solution formation which shows positive for multi-component alloys. ΔS_{mix} for AlCrFeCoNi as reported by researchers is 13.38 J/K mol which is much greater than the conventional alloys (0.22R–1.15R), the positive value can significantly lower ΔG_{mix} value promoting the random distribution of elements in the crystal lattice. The effect on ΔS_{mix} becomes more prominent with an increase in temperature in which atomic size difference plays an important role [15]. Since Al due to its smaller atomic radii as compared to Cr, Fe, Co, Ni elements can cause lattice distortion in the alloy which can also alter the compositional variation in the system. The reported values for all the parameters shown in Table 3 come in the range of high entropy alloy which confirms the high entropy alloy coatings formation.

3.4 Open Circuit Potential and Tafel Polarization Behavior

To study the corrosion behavior of mild steel, HEA_1 and HEA_2, EIS, and Tafel polarization measurements were carried out. To attain constant potential value, the coatings were immersed in 3.5% NaCl corrosive media (close to seawater concentration) for 1 h (Fig. 8a). The exposed area of the uncoated and coated samples was taken as 1 cm^2 for the study. The Tafel polarization was studied in the potential range of ± 300 mV which can be observed in the Fig. 8b. Furthermore, calculations were done to obtain parameters such as Ecorr (corrosion potential), icorr (corrosion current) density, cathodic and anodic slopes, and (R_p) polarization resistance value as tabulated in Table 4. Polarization resistance and corrosion resistance (mpy) in Tafel curves can be calculated using the following equations [16]:

$$R_p = \frac{\beta a + \beta c}{i_{\text{corr}}(\beta a \cdot \beta c)\ln 10} \tag{2}$$

$$CR\,(\text{in mpy}) = \frac{0.129 * i_{\text{corr}} * EW}{dA_{\text{ex}}} \tag{3}$$

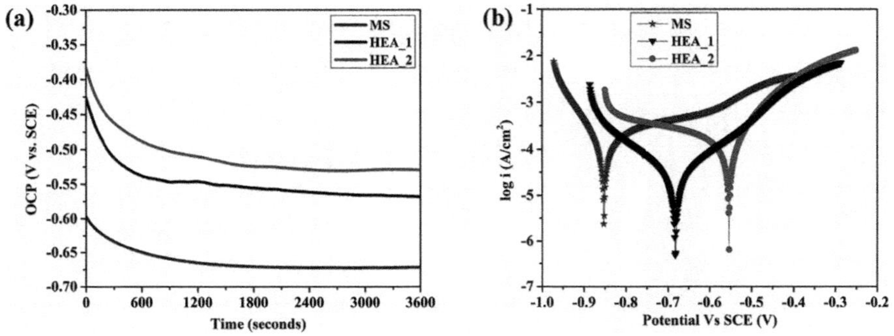

Fig. 8 **a** Open circuit potential (OCP) and **b** Tafel polarization curves for HEA_1 and HEA_2 samples

Table 4 Parameters obtained from extrapolation of Tafel curves for HEA_1 and HEA_2 samples

System	i_{corr} (μA/ cm^2)	E_{corr} (V)	Anodic slope (1/βa) (1/V)	Cathodic slope (1/ βc) (1/V)	Linear polarization (R_p) (Ohm)	P_{eff} (%) $\{\frac{i_{MS}-i_{HEA}}{i_{MS}}\}$	Corrosion rate (mpy)
MS	135.4 ± 4.5	−0.853	4.93	4.50	173.4	–	0.51
HEA_1	75.2 ± 2.4	−0.683	3.89	8.61	429.1	44.46	0.28
HEA_2	31.75 ± 2.6	−0.554	3.43	6.62	1049.7	76.55	0.12

Here, R_p: polarization resistance (Ω) i_{corr}: corrosion current density (μA/cm^2); βa: anodic slope; βc: cathodic slope; EW: equivalent weight (g/mol); d: density (g/cm^3); A_{ex}: exposed area (cm^2). The generated results in Table 4 showed that a positive shift in Ecorr value and decrease in icorr value from mild steel substrate to HEA_2 implies inertness of coating [12]. The anodic and cathodic slope, Tafel constant are not equal ($|\beta a| \neq |\beta c|$) forming a non-symmetric Tafel curve which indicates metal corrosion. The comparative larger value from Table 4 of $|\beta a|$ than $|\beta c|$ suggests the occurrence of corrosion because of anodic polarization [16]. Hence, the effect of coating and annealing on mild steel exhibits higher $|\beta a|$ values than $|\beta c|$. The decrease in corrosion current density (i_{corr}) was observed in HEA_1 (75.2 ± 2.4 μm) from substrate MS (135.4 ± 4.5 μm). The annealing furthermore decreased the corrosion current density value of HEA_2 (31.75 ± 2.6 μm). Figure 9 illustrates the plot of icorr and protection efficiency. The protection efficiency can be calculated using the formula (P_{eff} = iMS − iHEA/iMS), where iMS and iHEA are the corrosion current density of the substrate and HEA_1/HEA_2. The highest protection efficiency of 76.55% was observed for HEA_2 with respect to MS suggesting an increase in corrosion resistance. Also, the observations conclude improvement in corrosion-resistant property post annealing treatment. The decrease in corrosion rate was observed at nearly 44.5% in HEA_1, while for the post annealed sample (HEA_2), it was 77%. However, all the observations indicate that (a) the AlCrFeCoNi-HEA coatings enhance the corrosion-resistant

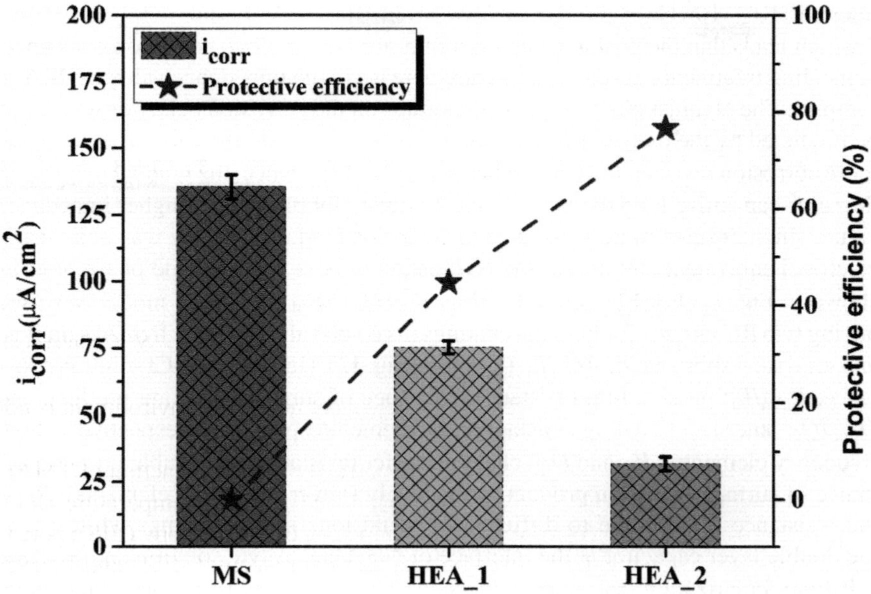

Fig. 9 Corrosion current density (i_{corr}) versus protective efficiency plot for MS, HEA_1, and HEA_2

property and (b) the annealing process furthermore improved the corrosion-resistant properties indicating better performance.

3.5 Electrochemical Impedance Spectroscopy (EIS) Study

EIS measurements were performed between 10^{-2} and 10^{6} Hz at 5 mV of sinusoidal amplitude after achieving constant OCP value for 1 h in 3.5% NaCl solution. Figure 10 illustrates the Nyquist plot for HEA_1 and HEA_2. Bode plots are shown in Fig. 11. Bode plots give information about the surface uniformity and compaction of the coatings in corrosive media. At lower frequencies, mod |Z| values correspond to the corrosion performance of the coatings; the higher the |Z| value, the more will be the corrosion resistance. The |Z| impedance values obtained were 338.42 Ω cm^2, 419.26 Ω cm^2, and 617.90 Ω cm^2 for MS, HEA_1, and HEA_2, respectively. The higher value of |Z| impedance values show a more protective and compact coating. The Bode phase plots showed two-time constants of the curves obtained between phase angles with frequency. The high-frequency loop denotes the passive film resistance, whereas the loop obtained at a higher frequency signifies the double-layer capacitive behavior. The upward trend of phase angle observed in HEA coating and post annealing process signifies improved homogeneity in the surface as reported by researchers [16]. The observed phase angle from Fig. 11 at a frequency range of 10^{-2}–10^{2} Hz is 35.46° for MS, 51.22° for HEA_1, and 57.09° for HEA_2 which leads that the post annealed coatings are less prone to corrosion resistance. Annealing treatment caused a significant increase in the impedance value in HEA_2 samples. The Nyquist plot in Fig. 10 elaborates on the corrosion behavior which can be predicted by the diameter of the loop formed; the greater the capacitive loop, the more corrosion resistant the coating becomes [12, 17]. Hence, the gradual increase in the semi-capacitive loop diameter of the Nyquist plot proposes a higher impedance value which ensures more resistance to corrosion [16]. Plot fitting was done using electrical equivalent circuits for the Nyquist, both bode |Z| and bode phase plots as shown in Fig. 11. For EEC circuit fitting, $R_{sol}(Q_{film}(R_{film}(Q_{dl}R_{ct})))$ model was used having two RC circuits for both the coatings (as coated and post heat treated samples) and mild steel substrate $R_{sol}(Q_f(R_f(Q_{dl}R_{ct})))$ (Fig. 12). Here, R_{sol}: NaCl solution resistance, R_{film}/R_f: passive film resistance/resistance of surface corrosion product, and Q_{film}/Q_f: outer layer coating constant phase element/capacitance, respectively (high frequency elements), R_{ct} and Q_{dl}: charge transfer resistance and double layer capacitance of surface corrosion product, respectively (lower frequency elements). R_{ct} is the resistance offered due to diffusion of metal ions in the coatings while Q_{dl} is the double layer capacitor at the interface of NaCl electrolyte solution and working substrate. Polarization resistance is the summation of all the resistance obtained by the coating; $R_p = R_{film} + R_{ct}$. A higher value of R_p suggests more corrosion resistance. The value of R_p increased from 302.13 Ω cm^2 (MS) to 619.33 Ω cm^2 (HEA_1) with a further increase to 1141.36 Ω cm^2 for HEA_2. As the coating surface is inhomogeneous, instead of using pure capacitance, CPE (constant phase element) is used. Z_Q is the impedance when constant phase element CPE is used which can be written as [7]:

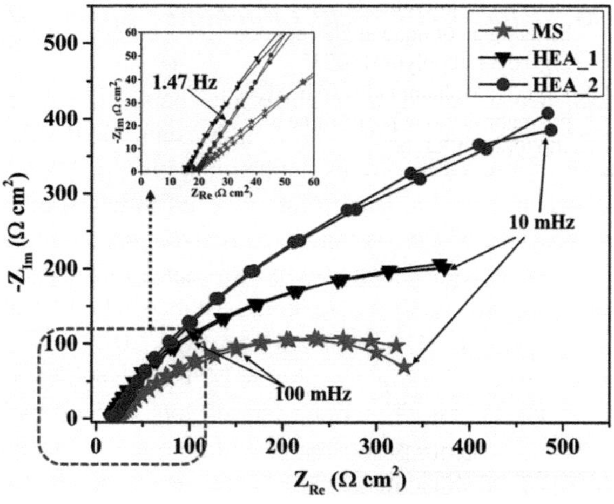

Fig. 10 Nyquist plots of MS, HEA_1, and HEA_2 samples

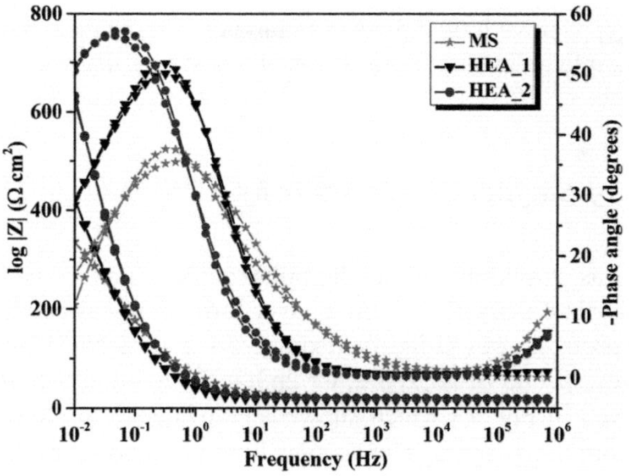

Fig. 11 Bode |Z| and bode phase plots of MS, HEA_1, and HEA_2 samples

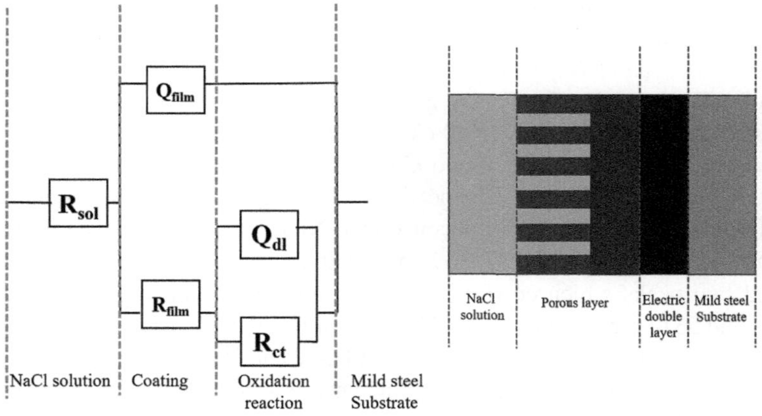

Fig. 12 Electrical equivalent circuits used for fitting in Nyquist curves fitted for HEA_1 and HEA_2 samples

$$Z_Q = \frac{1}{(j\omega)^n Y_0} \tag{4}$$

where ω: angular frequency (rad/s), Y_0: coefficient of non-ideal capacitive behavior, n: index of the corresponding constant phase element. The index varies as $n = -1$ for the ideal inductor, $n = 0$ for the ideal resistor, and $n = 1$ for the ideal capacitor. All the parameters R_{sol}, Q_{film}, R_{film}, Q_{dl}, R_{ct}, and Chi-squared (χ^2) obtained after fitting through electrical equivalent circuits are shown in Table 5. Replacing the parameters Q_{film} and R_{film} as Q_{f} and R_{f} for mild steel. The fitting curve was fitted with a fitness value of $\leq 10^{-3}$. The HEA coating led to more corrosion-resistant properties of the mild steel substrate. Also, HEA_2 (post annealed) showed higher value of |Z|, phase angle and semi capacitive loop diameter, polarization resistance than that of HEA_1 (as coated) sample indicating more corrosion resistance in corrosive media.

3.6 Nature of Surface Oxide Study by XPS

The corrosion was caused due to anodic polarization, as observed in Tafel curves (Fig. 8b). The further analysis of passive layer formation was carried out by X-ray photoelectron spectroscopy (XPS) after exposure to a 3.5% NaCl corrosive environment. The survey spectra of Al 2p3/2, Cr 2p3/2, Fe 2p3/2, Co 2p3/2, and Ni 2p3/2 can be observed in Fig. 13 which showed the evidence of elements present on the surface of both the samples. The deconvolution of peaks was done to study the high-resolution spectra of corresponding signals of Al 2p3/2, Cr 2p3/2, Fe 2p3/2, Co 2p3/2, Ni 2p3/2, O1s, and C1s. The deconvoluted peaks corresponding to fitting curves were plotted as shown in Figs. 14 and 15. For HEA_1: Al 2p3/2 corresponds to Al^0 and Al_2O_3 with binding energies of 73.96 and 75.32 eV, respectively [18]. Cr 2p3/2 was deconvoluted to CrO, $Cr(OH)_2$, and CrO_3 with binding energies of 573.89, 576.58, and 578.76 eV, respectively [12]. Fe 2p3/2 was split into Fe^0, FeO or Fe_3O_4, Fe_3O_4 or Fe_2O_3, Fe_2O_3, and FeOOH with binding energies of 706.07, 708.29, 711.23, 720.12, and 715.40 eV, respectively [19]. Co 2p3/2 corresponding peaks were metallic Co^0, CoO, and Co_3O_4 with corresponding binding energies 780.86, 784.99, and 788.81 eV, respectively [19]. The NiO, $Ni(OH)_2$, metallic Ni^0, and $Ni(OH)_2$ were observed by deconvoluted Ni2p3/2 spectra with corresponding binding energies 853.43, 856.27, 859.36, and 862.29 eV, respectively [19]. The O1s showed two peaks which correspond to O^{2-} and OH^- with binding energies of 530.72 and 531.75 eV, respectively. The C1s splits into C–C, $CO_3{}^{2-}$, and C=O with corresponding binding energies of 283.80, 284.48, and 287.31 eV, respectively [20]. For HEA_2, the deconvoluted peaks of elements for sample HEA_2 can be observed in Fig. 15. Al $2p_{3/2}$ splits into Al^0, Al_2O_3 with corresponding binding energies of 73.98 and 74.78 eV. A separate peak was observed for Al as $Al(OH)_2$ at binding energy 78.32 eV [21]. Cr $2p_{3/2}$ showed Cr^0, $Cr(OH)_3$, and CrO_3 with binding energies 573.58, 575.07, and 577.04 eV, respectively. Cr also exhibited Cr_2O_3 at a binding energy of 579.46 eV [12]. Fe $2p_{3/2}$ correspond to Fe^0, FeO or Fe_3O_4, Fe_3O_4 or Fe_2O_3 and Fe_2O_3 with binding energies 704.82, 706.56, 710.99 and 719.69 eV respectively [19]. Ni $2p_{3/2}$ splits into NiO and Ni^0 at binding energies 853.59 and 859.97 eV. Two peaks were obtained for $Ni(OH)_2$ at binding energies of 856.69 and 863.16 eV [19]. The O1s showed O^{2-} and OH^- with the presence of H_2O at binding energies of 530.32, 531.24, and 531.68 eV, respectively. The C1s splits into C–C, $CO_3{}^{2-}$, and C=O with binding energies of 283.94, 284.73, and 287.57 eV, respectively [20]. The binding energies can be observed as tabulated in Table 6.

Table 5 EEC fitted value for Nyquist for AlCrFeCoNi HEA as coated and post annealed in 3.5% NaCl corrosion media

System	R_{Sol} (Ω cm²)	n_{film}	R_{film}/R_f (Ω cm²)	Q_{film}/Q_f(Ω^{-1} cm^{-2} S^{-n})	n_{dl}	R_{Ct} (Ω cm²)	CPE$_{dl}$ (Ω^{-1} cm^{-2} S^{-n})	R_p (Ω cm²)	$\chi 2 \times 10^{-3}$
MS	15.71	0.89	28.23	2.428×10^{-5}	0.69	273.9	0.0239	302.13	0.24
HEA_1	20.47	0.98	61.13	6.75×10^{-5}	0.72	558.2	0.01586	619.33	0.71
HEA_2	22.24	0.99	58.36	1.65×10^{-5}	0.69	1083	0.03567	1141.36	0.51

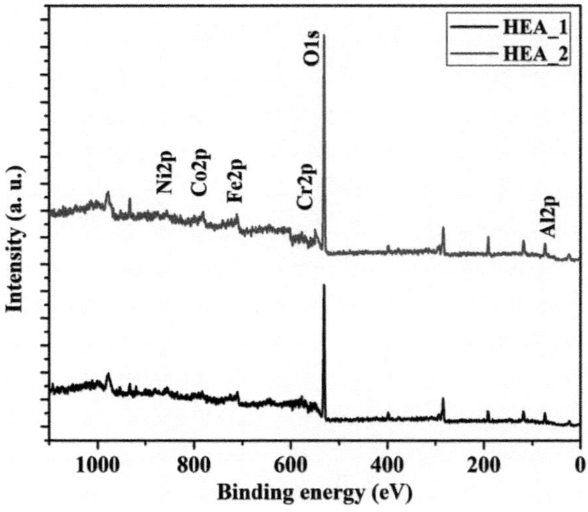

Fig. 13 XPS survey of HEA_1 and HEA_2 after corrosion

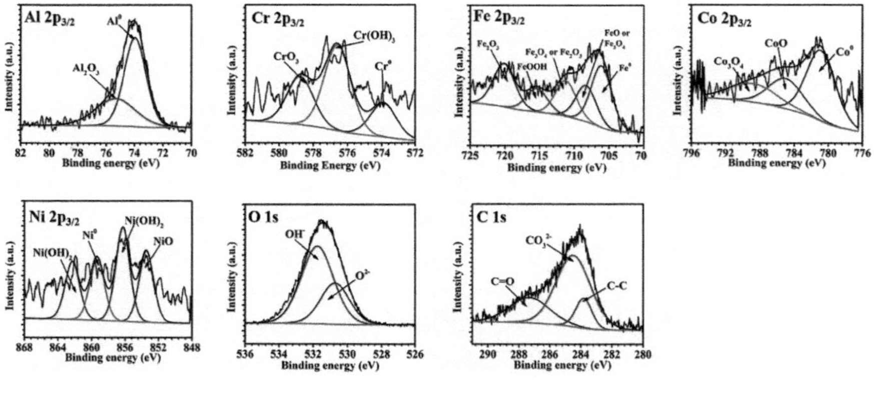

Fig. 14 High-resolution deconvoluted XPS spectra of HEA_1 sample after corrosion in 3.5% NaCl corrosive media

3.7 Relative Content Analysis: Cationic Fraction

The relative elemental content is tabulated in Table 6 which corresponds to various elemental states of passive films present. The area covering the photoelectron peaks is designated to the relative content percentage. Since the coatings were exposed to 3.5% NaCl corrosive media, more oxides and hydroxides formation was observed than the elemental states. In HEA_1, Al^0 and Co^0 were predominant elemental states along with Cr^{3+} ($Cr(OH)_3$) components with relative content of more than 50% each. FeOOH, an unstable hydroxide was also present in HEA_1 which was not observed later in HEA_2. The relative content of metallic elements decreased in HEA_2 as

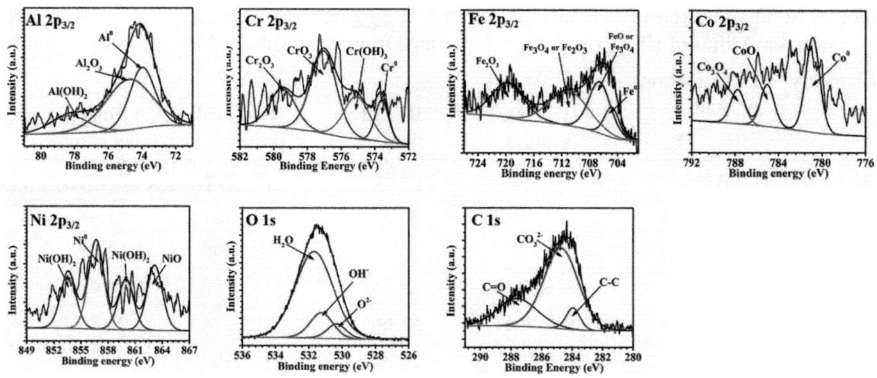

Fig. 15 High-resolution deconvoluted XPS spectra of HEA_2 sample after corrosion in 3.5% NaCl corrosive media

compared to HEA_1. Major oxides formed in HEA_2 were Co_3O_4 (Co_{ox}^{3+}), Al_2O_3 (Al_{ox}^{3+}), and Cr_2O_3 (Cr_{ox}^{3+}). Figure 16 shows the plotted relative content of all the elements, oxides, and hydroxides present on the passive layer.

The formation of extra peak of $Al(OH)_2$ in HEA_2 can be attributed to form more stable oxide films along with an increase in the relative content of Al_2O_3 from 35.34 eV (HEA_1) to 51.43 eV (HEA_2) [21]. The formation of Cr_2O_3 in HEA_2 has been observed which contributes to the corrosion-resistant properties in different systems as discussed by many authors [12] and the same was not present in the former. The enrichment of CrO_3 on the surface in HEA_2 suggests stable oxide film formation. The CrO_3/Cr^0 ratio was calculated for pristine and post annealed samples which was found to be almost three times higher for HEA_2 than HEA_1. The presence of $Cr(OH)_2$ in higher amounts in HEA_1 can lead to defective and porous structure increasing the unstability [12]. The cobalt is present in two oxidation states, Co^{2+} and Co^{3+} in both samples. Co_3O_4 is a more stable oxide for stabilizing the passive layer formation. The relative content of Co^{3+} (Co_3O_4) is dominant over Co^{2+} (CoO) which contributes to oxide layer stability [12].

The presence of Ni^{2+} ($Ni(OH)_2$) as a dominant oxide in HEA_2 than in HEA_1 was called upon by the slower rate of diffusion in the presence of other elements in the corrosive solution [19]. The absence of H_2O in HEA_ 1 but an additional peak in HEA_2 can also be correlated to the additional protective layer corresponding to bound water formation which traps the dissolved metal ions forming a barrier from corrosive media attack [22]. An increase in FeO cationic fractions suggests the robust passive oxide film because of its high density. The absence of FeOOH in HEA_2 decreases the overall hydroxide formation owing to stability. The oxide ratio $FeO/(Fe_2O_3+FeOOH)$ was 0.5 for HEA_1 and that increased for HEA_2 by two times. The higher $Fe_{ox}^{2+}/Fe_{ox}^{3+}$ improves the corrosion resistance behavior [7]. To higher corrosion resistant property has been attributed to the higher ratio of more corrosion resistant elements to the remaining less corrosive resistant elements [12]. The present analysis of the system forming (Al+Cr+Co)/(Fe+Ni) ratios gives higher

Table 6 Binding energies and relative content of Al $2p_{3/2}$, Cr $2p_{3/2}$, Fe $2p_{3/2}$, Co $2p_{3/2}$, and Ni $2p_{3/2}$ of passive films of HEA_1 and HEA_2 obtained from deconvoluted XPS spectra

Elements	Oxidation state	HEA_1		HEA_2	
		HEA peak position	Relative content	HEA peak position	Relative content
Al	Al^0	73.9636	64.66	73.9836	33.78
	Al_2O_3	75.3259	35.34	74.7851	51.43
	$Al(OH)_2$			78.3296	14.78
Cr	Cr^0	573.8930	19.76	573.5850	11.43
	$Cr(OH)_3$	576.5844	51.66	575.0774	22.87
	CrO_3	578.7661	28.58	577.0402	44.39
	Cr_2O_3			579.4602	21.31
Fe	Fe^0	706.0777	30	704.8269	10.86
	FeO or Fe_3O_4	708.2980	17.52	706.5666	27.85
	Fe_3O_4 or Fe_2O_3	711.2310	17.80	710.9917	33.79
	Fe_2O_3	720.1207	22.53	719.6937	27.50
	FeOOH	715.4049	12.15		
Co	Co^0	780.8692	51.78	780.8646	24.53
	CoO	784.9914	25.75	785.0200	20.04
	Co_3O_4	788.8181	22.48	787.7548	55.43
Ni	NiO	853.4385	24.95	853.5953	21.79
	$Ni(OH)_2$	856.2746	32.66	856.6919	34.07
	Ni^0	859.3600	22.12	859.9717	19.27
	$Ni(OH)_2$	862.2970	20.26	863.1670	24.87
C	C–C	283.8073	12.04	283.9454	7.91
	CO_3^{2-}	284.4812	62.54	284.7369	66.87
	C=O	287.3180	25.42	287.5741	25.94
O	O^{2-}	530.7231	31.91	530.3293	6.97
	OH^-	531.7564	68.09	531.2448	14.36
	H_20			531.6835	78.67

value for HEA_2 than HEA_1 which can be attributed to the fact to more corrosive resistant nature of HEA_2.

The relative content plotted data in decreasing sequence are as follows:

For HEA_1: $Al^0 > Co^0 > Cr(OH)_3 > Al_2O_3 > Ni(OH)_2 > Fe^0 > CrO_3 > CoO > NiO > Fe_2O_3 > Co_3O_4 > Ni^0 > Ni(OH)_2 > Cr^0 > Fe_3O_4$ or $Fe_2O_3 >$ FeO or $Fe_3O_4 >$ FeOOH.

For HEA_2: $Co_3O_4 > Al_2O_3 > CrO_3 > Fe_3O_4$ or $Fe_2O_3 > Al^0 > Ni(OH)_2 >$ FeO or $Fe_3O_4 > Fe_2O_3 > Ni(OH)_2 > Co^0 > Cr(OH)_3 > NiO > Cr_2O_3 > CoO > Ni^0 > Al(OH)_2 > Cr^0 > Fe^0$

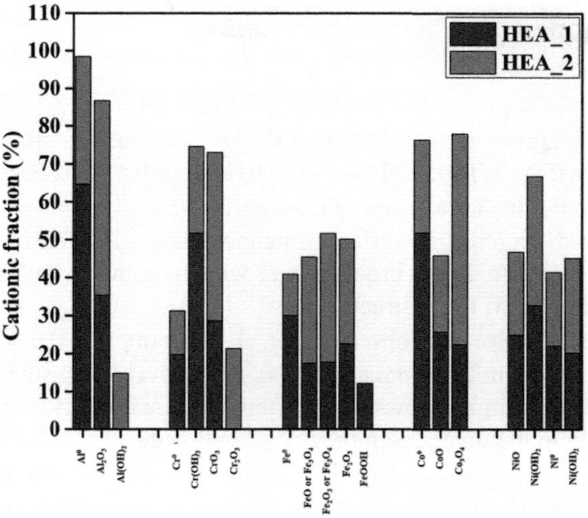

Fig. 16 Relative contents (cationic fractions) of the passive layer in HEA_1 and HEA_2 samples

Based on the relative contents and SEM micrograph of the cross-section of the coatings (Fig. 17). the oxide layer formation mechanism of HEA_1 and HEA_2 samples is depicted.

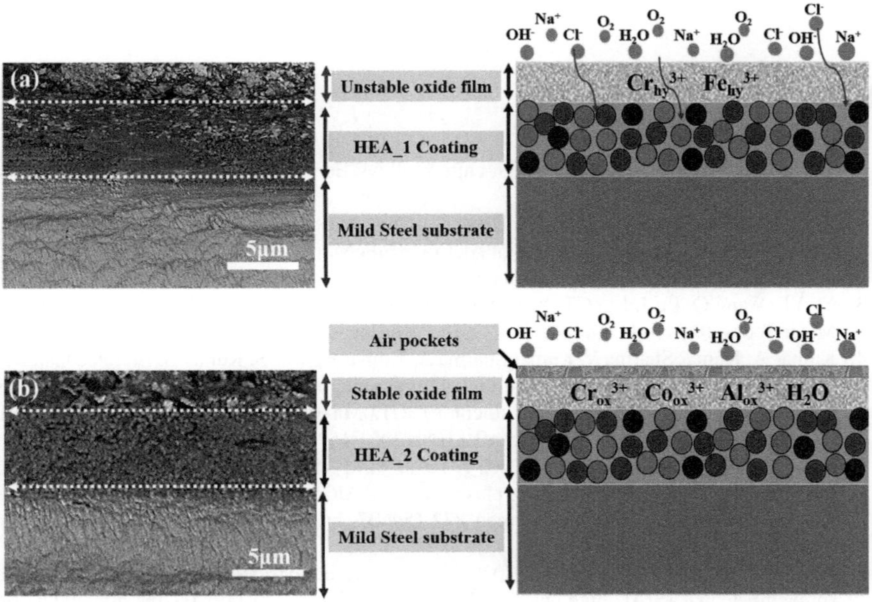

Fig. 17 SEM micrograph of coating thickness and oxide layer formation mechanism of **a** HEA_1 and **b** HEA_2 samples

4 Conclusions

1. The electroless deposition of AlCrFeCoNi HEA showed fine grained, compact coating which promotes uniformity of the surface coatings after post annealing treatment in HEA_2. The SEM micrographs of the HEA_2 sample were observed, which showed globular and smooth morphology.
2. The observed decrease in surface roughness from 1.218 μm for HEA_1 to 0.7112 μm for HEA_2 goes in accordance with the compact and smooth granules as observed by SEM microstructure.
3. BCC phase structure was observed for all the samples. HEA_1 and HEA_2 samples were rich in Ni-Cr-Fe and Al-Ni, respectively. The stable phase formation for both the samples shows the high-temperature stability for coatings despite the reduction in the microstrain in HEA_2 with annealing.
4. Higher polarization resistance (R_p) value (1141.36 Ω cm^2), lower corrosion current density (31.75 μA/cm^2) and decrease in corrosion rate up to 77% in HEA_2 sample as compared to MS substrate suggest improved corrosion resistance behavior of coatings in HEA_2 sample.
5. The major constituents of HEA_1 was found to be ${Co_{ox}}^{3+}$, ${Al_{ox}}^{3+}$, and ${Cr_{ox}}^{3+}$ which acts as protective coatings, while HEA_2 revealed higher hydroxides ${Cr_{hy}}^{3+}$ and ${Fe_{hy}}^{3+}$. The hydrophobic HEA_2 coating with a contact angle of 145.35° exhibited a more protective layer on the surface preventing chloride ions from penetrating through the coatings.

References

1. Yeh JW (2006) Recent progress in high-entropy alloys. European J Control 31:633–648. https://doi.org/10.3166/acsm.31.633-648
2. Xu XD, Guo S, Nieh TG, Liu CT, Hirata A, Chen MW (2019) Effects of mixing enthalpy and cooling rate on phase formation of Al_xCoCrCuFeNi high-entropy alloys. Materialia (Oxf) 6:100292. https://doi.org/10.1016/j.mtla.2019.100292
3. Ye YF, Wang Q, Lu J, Liu CT, Yang Y (2016) High-entropy alloy: challenges and prospects. Mater Today 19:349–362. https://doi.org/10.1016/j.mattod.2015.11.026
4. Kumar A, Kumar Sharma V, Kumar Sharma A, Kumar Biswas S, Kumar S, Muraleedharan K (2018) Effect of quenching and partitioning on the microstructure and mechanical properties of a medium carbon low alloy low silicon (0.3C1.0Mn–0.4Si–0.6Cr–0.46Ni–0.26Mo) steel. Mat Res Exp 5:116503. https://doi.org/10.1088/2053-1591/aadb7d
5. Wang L, Zhang F, Yan S, Yu G, Chen J, He J, Yin F (2021) Microstructure evolution and mechanical properties of atmosphere plasma sprayed AlCoCrFeNi high-entropy alloy coatings under post-annealing. J Alloys Compd 872:159607. https://doi.org/10.1016/j.jallcom.2021.159607
6. Xiao JK, Wu YQ, Chen J, Zhang C (2020) Microstructure and tribological properties of plasma sprayed FeCoNiCrSiAl_x high entropy alloy coatings. Wear 448–449:203209. https://doi.org/10.1016/j.wear.2020.203209
7. Singh S, Srivastava C (2023) Effect of carbon nanotube incorporation on the evolution of morphology, phase and compositional homogeneity, surface oxide chemistry and

corrosion behaviour of electrodeposited FeCuMnNiCo-carbon nanotube composite coatings. Electrochim Acta 439:141639. https://doi.org/10.1016/j.electacta.2022.141639

8. Aliyu A, Srivastava C (2019) Microstructure-corrosion property correlation in electrodeposited AlCrFeCoNiCu high entropy alloys-graphene oxide composite coatings. Thin Solid Films 686:137434. https://doi.org/10.1016/j.tsf.2019.137434
9. Dang C, Dong M, Li J (2019) Influence of annealing temperature on film morphology and tribological performance of TiSiN-Ag coating. Mater Res Express 6:096446. https://doi.org/10.1088/2053-1591/ab0d2f
10. Toloei A, Stoilov V, Northwood D (2013) The relationship between surface roughness and corrosion. ASME Proc: Advan Manuf 2B:65498. https://doi.org/10.1115/IMECE2013-65498
11. Zhang H, Pan Y, He Y (2011) Effects of annealing on the microstructure and properties of 6FeNiCoCrAlTiSi high-entropy alloy coating prepared by laser cladding. J Thermal Spray Technol 20:1049–1055. https://doi.org/10.1007/s11666-011-9626-0
12. Singh S, Srivastava C (2023) Engineering the protective oxide chemistry for enhanced corrosion protection performance of FeCuCrNiCo-CNT composite coatings in 3.5 M NaCl solution corrosive media. Metall Mater Trans A 54:1398–1413. https://doi.org/10.1007/s11661-023-06994-1
13. Wu ZX, He MJ, Feng CS, Wang TL, Lin MZ, Liao WB (2022) Effects of annealing on the microstructures and wear resistance of CoCrFeNiMn high-entropy alloy coatings. J Thermal Spray Technol 31:1244–1251. https://doi.org/10.1007/s11666-021-01292-x
14. Zheng T, Hu Y, Zhang Y, Pan F (2017) Formation of a hydrophobic and corrosion resistant coating on magnesium alloy via a one-step hydrothermal method. J Coll Interf Sci 505:87–95. https://doi.org/10.1016/j.jcis.2017.05.092
15. Yang X, Zhang Y (2012) Prediction of high-entropy stabilized solid-solution in multi-component alloys. Mater Chem Phys 132:233–238. https://doi.org/10.1016/j.matchemphys.2011.11.021
16. Jyotheender KS, Srivastava C (2019) Ni-graphene oxide composite coatings: optimum graphene oxide for enhanced corrosion resistance. Compos B Eng 175:107145. https://doi.org/10.1016/j.compositesb.2019.107145
17. Lala SRF, Gupta A, Srivastava C (2022) Texture evolution and corrosion behaviour of Sn–1.5 wt% Cr coatings containing graphene oxide. Philos Mag 102:522–541. https://doi.org/10.1080/14786435.2021.2011981
18. Digne M, Sautet P, Raybaud P, Toulhoat H, Artacho E (2002) Structure and stability of aluminum hydroxides: a theoretical study. J Phys Chem 106:5155–5162. https://doi.org/10.1021/jp014182a
19. Biesinger MC, Payne BP, Grosvenor AP, Lau LWM, Gerson AR, Smart RSC (2011) Resolving surface chemical states in XPS analysis of first row transition metals, oxides and hydroxides: Cr, Mn, Fe, Co and Ni. Appl Surf Sci 257:2717–2730. https://doi.org/10.1016/j.apsusc.2010.10.051
20. Chen X, Wang X, Fang D (2020) A review on C1s XPS-spectra for some kinds of carbon materials. Fuller Nanotub Carbon Nanostruct 28:1048–1058. https://doi.org/10.1080/1536383X.2020.1794851
21. Sherwood PMA (1998) Introduction to studies of aluminum and its compounds by XPS. Surf Sci Spectra 5:1–3. https://doi.org/10.1116/1.1247880
22. Singh S, Srivastava C (2023) On the evolution of protective oxide chemistry and corrosion resistance of electrodeposited FeCrMnNiCo-carbon nanotube composite coatings. Philos Mag 103:1–24. https://doi.org/10.1080/14786435.2022.2159091

The Corrosion Behaviour of a Pyrazoline Derivative on Mild Steel in Hydrochloric Acid Medium: Electrochemical and Quantum Chemical Investigation

Prathima Shekara, Jyothi Kudva, Rajitha Sadashiva, and Damodara Naral

1 Introduction

Most of the industries use iron and its alloys due to its magnificent physical, mechanical and other properties [1]. The harsh environment of the petrochemical industries can be endured by mild steel, an important alloy of iron. The corrosion of iron and its alloy is a global issue that affects the overall GDP of a nation. One feasible method to reduce the corrosion rate is to employ organic compounds as corrosion inhibitors [2]. These compounds preferably block redox reactions and are involved in the corrosion process. The organic compounds containing heterocycles, conjugated systems and heteroatoms participate in the donor–acceptor interaction between the metal and the inhibitor, thereby reducing the rate of metal dissolution [3]. Among the various heterocyclic organic compounds, pyrazoline derivatives are proven to be efficient organic inhibitors. These molecules are known to play multiple roles such as antimicrobial, antitubercular, anticancer and antifungal agents. Apart from possessing medicinal applications, the pyrazoline derivatives are also used as anticorrosive agents for iron, mild steel, copper and aluminium [4]. The present work involves the investigation of N-methyl-5-(4-methyl phenyl)-3-phenyl-4,5-dihydro-1H-pyrazole-1-carbothioamide (MPPC) as a corrosion inhibitor for mild steel in 0.5 M HCl. The electrochemical analysis is carried out by using electrochemical impedance spectroscopy and potentiodynamic studies. The surface analysis

P. Shekara · D. Naral
Department of Chemistry, St Joseph Engineering College, Vamanjoor, Karnataka, India

J. Kudva (✉) · D. Naral
Sigma-Aldrich Chemicals Pvt. Ltd., Bengaluru, Karnataka, India
e-mail: jyothik@sjec.ac.in

R. Sadashiva · D. Naral
Alvas Institute of Technology, Moodabidre, Karnataka, India

S. T. Aruna et al. (eds.), *Recent Advances in Electrochemical Science and Technology*, Springer Proceedings in Materials 47, https://doi.org/10.1007/978-981-97-3958-5_12

Table 1 Composition of various elements present in the mild steel sample

Elements	Fe	C	Mn	P	Al	Si	S
Atomic, composition (% weight)	99.6	0.24	0.007	0.02	0.02	0.04	0.01

was carried out by scanning electron microscopy and EDX analysis. The theoretical confirmation for the experimental study was provided by quantum chemical calculations.

2 Materials and Methods

2.1 *Material Preparation*

The mild steel used for the corrosion inhibition study is composed of various elements and its atomic composition is given in Table 1.

2.2 *Preparation of Corrosive Medium and Inhibitor Solution*

As hydrochloric acid is used as a pickling agent in various industries, for the present study 0.5 M HCl was used as the corrosive medium. The corrosion-inhibiting solutions were composed of various concentrations of MPPC (concentrations ranging from 5 to 50 ppm).

2.3 *Electrochemical Techniques*

A three-electrode system potentiostat (Model: CHI608 E-series) was used to perform electrochemical studies. For electrochemical impedance studies, a frequency range of 10,000–0.01 Hz, and for potentiodynamic polarization, a potential range of −250 to +250 mV at a sweep rate of 0.01 mV was employed.

2.4 *Surface Analysis*

The surface morphology of the specimen in uninhibited and inhibited media is analysed by scanning electron microscopy (Carl Zeiss FE-SEM model) and energy-dispersive X-ray spectroscopy.

2.5 Quantum Chemical Calculation

The theoretical quantum calculations were carried out with the help of the DFT/B3LYP level using Gaussian 16 software.

3 Results and Discussion

3.1 Electrochemical Impedance Studies

The electrochemical impedance spectroscopy was employed for different concentrations of MPPC (5–50 ppm) with varying temperatures (303–323 K) in 0.5 M HCl medium. The impedance spectra obtained were further analyzed using Z-simpwin software to check the variation in resistance of the specimen during the charge transfer reaction. The frequency range of 100 kHz to 10 MHz with an amplitude of 5 mV was chosen for the electrochemical impedance measurements.

Though the Nyquist plot (Fig. 1) of blank and inhibited media looks similar, there is a consistent widening of the impedance plot with an increase in MPPC concentration. These widened semicircles evince the increase in charge transfer resistance thereby confirming the corrosion inhibitory action of MPPC [5]. It is also observed that the impedance plot of the inhibited solution is not as smooth as the blank which is due to the adsorption or desorption of MPPC on the uneven surface of the metal [6]. The simple Randle's circuit (Fig. 2) was used for the EIS data analysis.

The constant phase element (CPE) was calculated with the help of capacitance (Y_0), the constant j and angular frequency (ω) as in Eq. 1.

$$Z_{CPE} = \frac{1}{Y_0(jw)^n} \tag{1}$$

The double-layer capacitance C_{dl} calculation was performed using Eq. 2.

$$C_{dl} = [Y_0 R_{ct}^{(1-n)}]^{\frac{1}{n}} \tag{2}$$

The corrosion inhibition efficiency, which describes the potential of MPPC, is calculated using Eq. 3

$$\%\ IE = \frac{R_p - R_p^0 \times 100}{R_p} \tag{3}$$

It is clear from Table 2 that there is a considerable increase in the inhibition efficiency with an increase in the concentration of MPPC and maximum efficiency of 95.1% was recorded in 50 ppm of MPPC at 303 K.

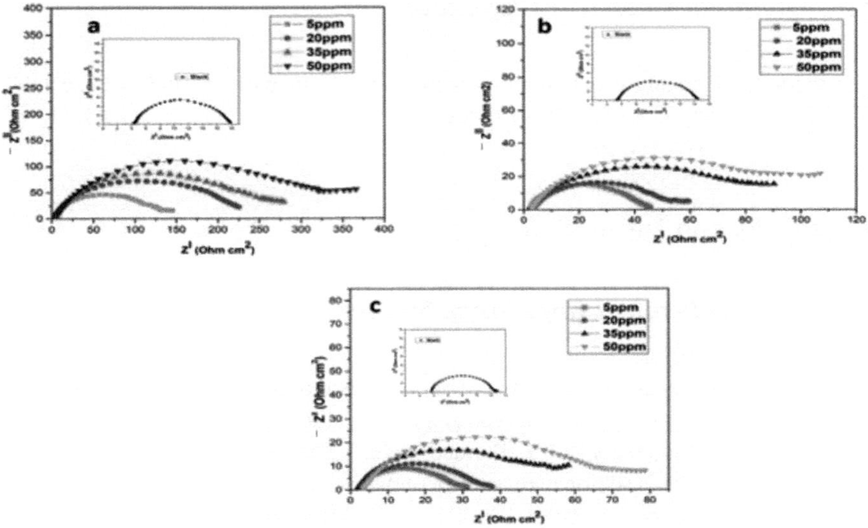

Fig. 1 **a** Nyquist plots for mild steel in 0.5 M HCl solution containing varying concentrations of MPPC at **a** 303 K, **b** 313 K, **c** 323 K

Fig. 2 The equivalent circuit model used for the fitting of impedance data

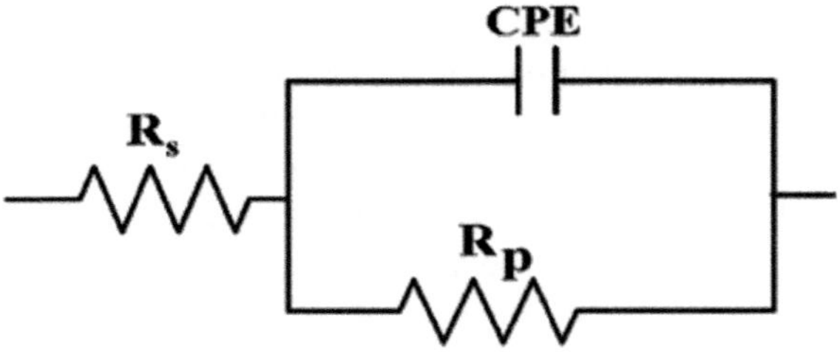

3.2 Potentiodynamic Polarization Studies

This study helps to realize the variation of corrosion current during the inhibitory action of MPPC on test metal. Figure 3 reveals the widely spaced polarization curves and negative shift of corrosion potential in the inhibited medium. This observation confirms the inhibitive action of MPPC on the cathodic reaction by reducing the rate of hydrogen evolution [7]. The corrosion inhibition was calculated by the values of corrosion currents of uninhibited (i^0_{corr}) and inhibited (i_{corr}) solutions as given in Eq. 4

$$IE = \frac{i^0_{\mathrm{corr}} - i_{\mathrm{corr}}}{i^0_{\mathrm{corr}}} \tag{4}$$

Table 3 shows the escalation of corrosion inhibition with successive hikes in MPPC concentration. As the displacement of E_{corr} is less (lesser than ±85 mV),

Table 2 The impedance data obtained for MPPC for mild steel

Temperature (°C)	Concentration (ppm)	R_s (Ω cm²)	R_{ct} (Ω cm²)	C_{dl} (μF/cm²)	n	% IE
30	Blank	2.9	12.73	303.2	0.846	
	5	2.3	136.3	90.59	0.751	90.66
	20	1.238	151.2	78.37	0.753	91.58
	35	2.031	217.9	56.8	0.784	94.15
	50	2.682	263.8	42.89	0.839	95.17
40	Blank	1.79	9.943	404.9	0.84	
	5	2.32	43.28	199.5	0.718	77.04
	20	1.909	36.29	112.8	0.745	72.57
	35	1.907	83.07	81.2	0.708	88.03
	50	3.464	63.92	61.5	0.767	84.44
50	Blank	3.807	9.53	215.6	0.86	
	5	2.587	28.69	89.8	0.8	66.78
	20	1.81	54.28	98.18	0.71	70.49
	35	2.599	54.13	64.48	0.8	82.42
	50	2.587	49.33	39.94	0.78	80.113

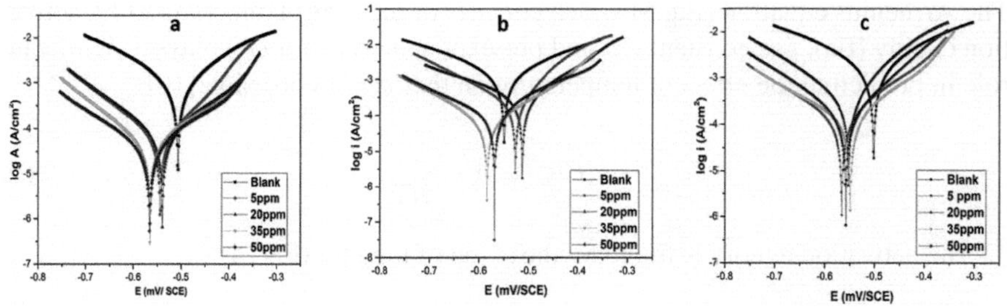

Fig. 3 Potentiodynamic polarization plot obtained during the corrosion of mild steel in 0.5 M HCl containing different concentrations of MPPC at **a** 303 K, **b** 313 K and **c** at 323 K

MPPC can be categorized under mixed-type inhibitor. The maximum efficiency of 97.6% is obtained in 50 ppm of MPPC at room temperature [8].

3.3 *Effect of Temperature on Inhibition*

The data obtained from electrochemical studies confirm the deceleration of corrosion inhibition efficiency with an increase in temperature. This trend reveals the physisorption behaviour of MPPC, which gets desorbed at elevated temperatures [9].

Table 3 Potentiodynamic polarization data at different concentrations of MPPC

Temperature	Concentration (ppm)	$-E_{corr}$ (mV/ SCE)	I_{corr} (uA cm^{-2})	βc	βa	n_p (%)
30 °C	Blank	−0.517	1122	158	147	
	5	−0.514	76.8	130	87	93.15
	20	−0.518	60.18	133	129	94.63
	35	−0.515	46.9	122	131	95.81
	50	−0.522	26.9	134	220	97.60
40 °C	Blank	−0.512	1282	187	177	
	5	−0.522	320	137	73	75.01
	20	−0.527	150.5	199	109	88.3
	35	−0.505	109.6	129	141	91.45
	50	−0.509	68.8	193	115	94.63
50 °C	Blank	−0.509	1381	173	154	
	5	−0.516	338	147	100	75.52
	20	−0.516	203.8	171	135	85.24
	35	−0.516	151.2	130	101	89.05
	50	−0.517	146.6	159	108	89.38

The Arrhenius equation (Eq. 5) which consists of the terms temperature (T), activation energy (Ea), gas constant (R) and pre-exponential factor (A) plays a significant role in predicting the effect of temperature on the rate of corrosion [10]

$$\ln(CR) = A - \frac{E_a}{RT} \tag{5}$$

The activation energy is found by the slope of the plot (Fig. 4).

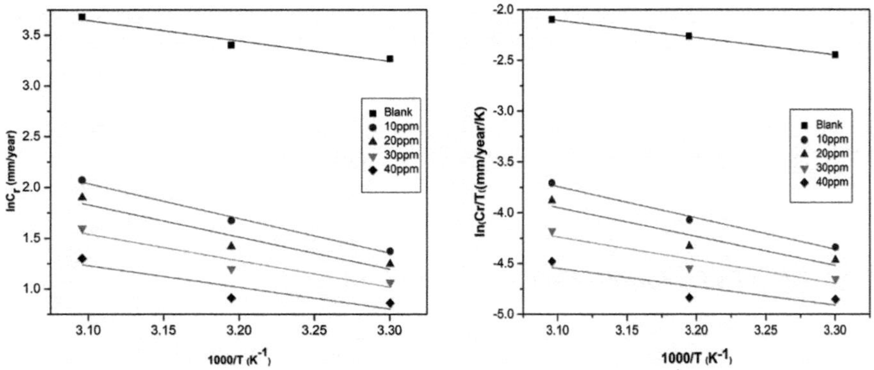

Fig. 4 **a** Arrhenius plot and **b** Arrhenius transition plot

Table 4 Activation parameters for the mild steel in MPPC

Inhibitor concentration (ppm)	Ea (kJ/mol)	ΔH (kJ/mol)	ΔS (J/mol K)
Blank	8.47	5.87	−192.53
10	51.09	57.94	−54.46
20	64.12	61.44	−45.45
30	55.62	45.72	−73.89
40	69.01	58.39	−62.53

The other activation parameters like entropy change ($\Delta S^{\#}$) and enthalpy change ($\Delta H^{\#}$), were determined by the Arrhenius transition equation (Eq. 6), where N and h represent Avagadro number and planks constant, respectively

$$\mathrm{CR} = \frac{\mathrm{RT}}{\mathrm{Nh}} \exp\left(\frac{\Delta \mathrm{S}^{\#}}{\mathrm{R}}\right) \exp\left(\frac{-\Delta \mathrm{H}^{\#}}{\mathrm{RT}}\right) \tag{6}$$

The higher and positive activation energy of the inhibited solution (Table 4) confirms the physisorption of MPPC being the endothermic mode of interaction, thereby arresting the rate of metal dissolution.

3.4 Adsorption Isotherms

As the inhibition of corrosion is mainly associated with the adsorption of the inhibitor on a metal surface, it is essential to test the adsorption isotherms for the given system to get clarity about the adsorption mechanism [11]. The present study fits in Langmuir adsorption isotherm, which is plotted using the concentration of the inhibitor (c) against the surface coverage (θ) (Eq. 7)

$$\frac{C}{\theta} = \frac{1}{K_{\mathrm{ads}}} + C \tag{7}$$

The standard adsorption free energy ($\Delta G^0_{\mathrm{ads}}$) is calculated (Eq. 8) using the coefficient of adsorption as obtained from the plot (Fig. 5)

$$K_{\mathrm{ads}} = \frac{1}{55.5}\left(\frac{-\Delta G^0_{\mathrm{ads}}}{RT}\right) \tag{8}$$

The negative values of $\Delta G^0_{\mathrm{ads}}$ (Table 5) validate the adsorption of MPPC as a spontaneous process. The $\Delta G^0_{\mathrm{ads}}$ value lesser than -40 kJ mol^{-1} shows mixed-type adsorption confirming the strong interaction between MPPC and mild steel surface [12].

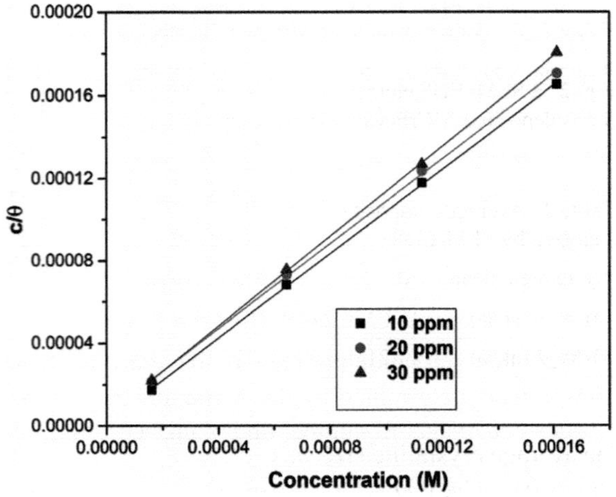

Fig. 5 Langmuir isotherm plot for the mild steel in 0.5 M HCl with BMDP

Table 5 Adsorption parameters obtained for mild steel corrosion in 0.5 M HCl containing BMDP at various temperatures

Temperature (K)	Slope	K_{ads}, (L/g)	ΔG^0 (kJ/mol)
303	0.999	597,950.2	−37.8
313	0.998	167,504.2	−35.7
323	0.999	214,592.3	−37.5

3.5 *Quantum Chemical Calculations*

The quantum chemical studies link the structure and anticorrosive properties of an inhibitor. It also verifies the experimental data with the theoretical data.

The optimized structures, HOMO and LUMO electronic distributions of MPPC, are given in Fig. 6. As observed in the table, the HOMO energy (E_{HOMO}) is high and LUMO energy (E_{LUMO}) is low which validates the association of electron-rich frontier molecular orbitals with corrosion prevention [13].

The parameters involved in the theoretical calculations were calculated and given in Table 6. The considerably less energy gap (ΔE) implies the tendency of HOMO to offer electrons to LUMO. The chemical rigidity and reactiveness of MPPC are given by hardness and softness, respectively. Noticeably low electronegativity and a high fraction of electrons prove MPPC as a highly reactive inhibitor [14].

3.6 *Surface Analysis*

The SEM micrographs are helpful in analyzing the surface changes in inhibited and uninhibited media. These images are usually taken 24 h after exposure in the presence and absence of MPPC.

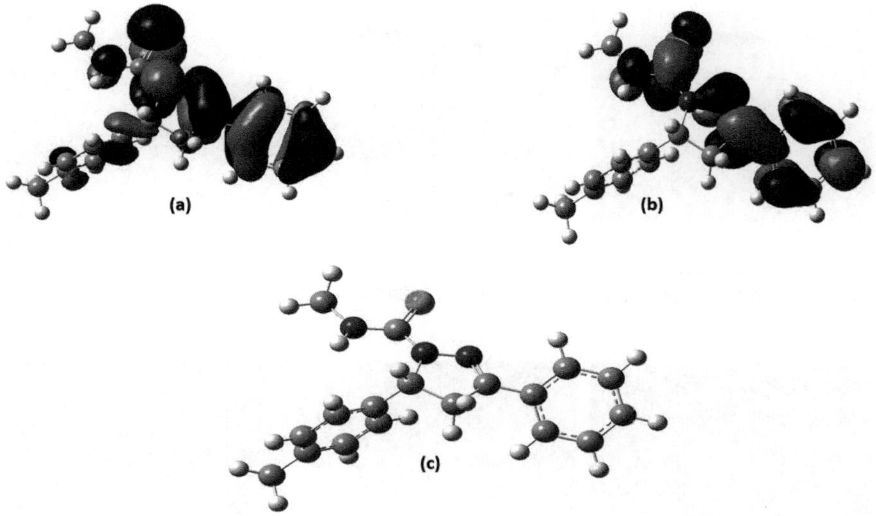

Fig. 6 **a** HOMO-frontier orbital, **b** LUMO-frontier orbital, **c** optimized structure of neutral MPPC

Table 6 Frontier molecular orbital data by Gaussian software

Parameters (B3LYP method)	Values (eV) (Unprotonated MPPC)
Energy of HOMO	−6.504
Energy of LUMO	−1.897
ΔE	4.607
– E_{HOMO} (I)	6.504
– E_{LUMO} (A)	1.897
Electronegativity(χ)	4.2005
Hardness (η)	2.303
Softness (σ)	0.435
ΔN (fraction of electrons transferred)	0.6078

It is evident from Fig. 7 that inhibited images are smoother than blank, proving the formation of the protective barrier by MPPC on the metal surface.

The EDX studies confirm the retention of iron content in the presence of MPPC. Table 7 shows the presence of a high amount of carbon and the appearance of a nitrogen peak in the inhibited spectrum which authenticates the role of various functional groups of MPPC in the adsorption process [15].

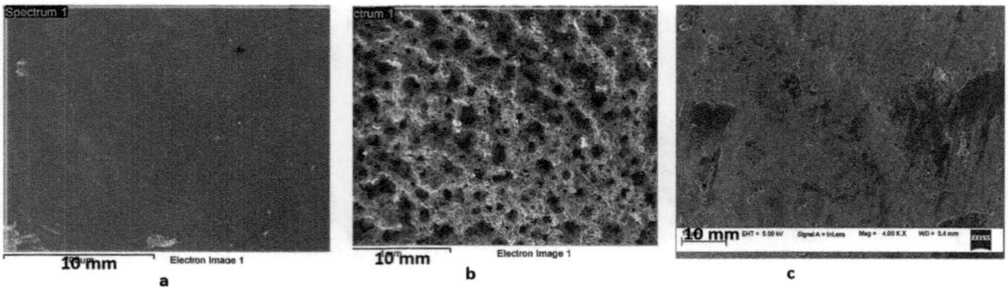

Fig. 7 SEM micrographs of **a** polished, **b** blank, **c** inhibited mild steel specimens

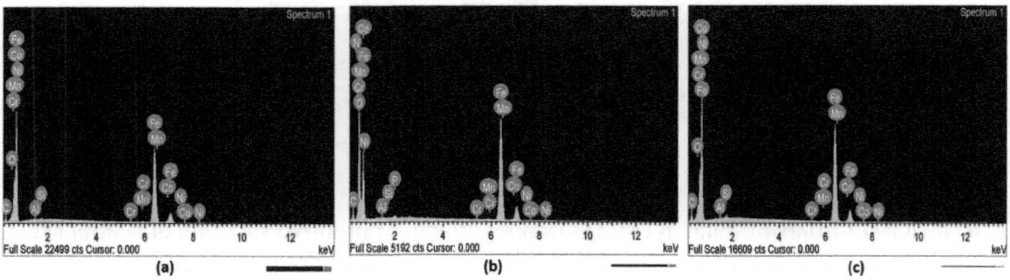

Fig. 8 EDX spectra of **a** polished, **b** blank and **c** inhibited mild steel specimen

Table 7 The atomic composition of the specimen by EDX analysis

Medium	Atomic composition [% weight]				
	O	Fe	Mn	C	N
Polished mild steel	4.28	91.66	0.72	2.14	–
Mild steel (in 0.5 M HCl)	44.2	31.61	0.21	4.72	–
Mild steel (in 0.5 M HCl with MPPC)	7.77	81.73	0.61	8.1	1.79

3.7 Comparison of Inhibition Efficiency with Similar Reported Compounds

The literature review on electrochemical studies of pyrazoline derivatives confirms that MPPC has greater potential to arrest corrosion compared to reported similar compounds (Table 8).

Table 8 The comparison of inhibition efficiency of MPPC with similar reported compounds

S. no	Name of the compound	Inhibitor concentration (ML^{-1})	Inhibition efficiency (%)	
			EIS	Tafel
	Pyrazolines containing carbothiamide derivatives [8]	10^{-3}	68	–
2	Pyrazolines possessing pyridine and methyle groups [16]	10^{-4}	83	82
3	Pyrazoles with carboxylate derivatives [17]	10^{-4}	83	83
4	Dimethylamine derivatives of pyrazolines [18]	10^{-4}	64	82
5	Pyrazolines with bisphenol moities [19]	10^{-4}	84	86
6	**MPPC (present work)**	$\mathbf{10^{-4}}$	**95**	**97**

4 Conclusion

The new pyrazoline derivative MPPC was tested as a corrosion inhibitor for mild steel in hydrochloric acid media. The electrochemical impedance spectroscopy, potentiodynamic polarization studies, surface analysis studies and quantum chemical calculations were used as the experimental tools to validate MPPC as a corrosion inhibitor. The experimental results have confirmed MPPC as a mixed type of inhibitor. The activation parameters and adsorption isotherms reveal spontaneous adsorption of MPPC on metal surfaces. The good agreement between the experimental studies confirms MPPC as a potential anticorrosive molecule [20, 21].

References

1. Lgaz H, Salghi R, Chaouiki A, Shubhalaxmi SJ, Bhat SK (2018) Pyrazoline derivatives as possible corrosion inhibitors for mild steel in acidic media: a combined experimental and theoretical approach. Cogent Eng. 5:1–17. https://doi.org/10.1080/23311916.2018.1441585
2. Sehmi A, Ouici HB, Guendouzi A, Benali FM, Boudjellal F (2020) Corrosion inhibition of mild steel by newly synthesized pyrazole carboxamide derivatives in HCl acid medium: experimental and theoretical studies. J Electrochem Soc 15:167. https://doi.org/10.1149/1945-7111/abab25
3. Anusuya N, Sounthari P, Saranya J, Parameswari K, Chitra S (2015) Corrosion inhibition effect of hydroxy pyrazoline derivatives on mild steel in sulphuric acid solution together with quantum chemical studies. J Mater Environ Sci 6:1606–1623
4. Abbas MA, Eid AM, Abdou MM, Elgendy A, Saeed RAE, Zaki EG (2021) Multifunctional aspects of the synthesized pyrazoline derivatives for AP1 5L X60 steel protection against

MIC and acidization: electrochemical, in silico, and SRB insights. ACS Omega 6:8894–8907. https://doi.org/10.1021/acsomega.0c06050
5. Siğircik G, Tüken T, Erbil M (2016) Assessment of the inhibition efficiency of 3,4-diaminobenzonitrile against the corrosion of steel. Corros Sci 102:437–445. https://doi.org/10.1016/j.corsci.2015.10.036
6. Tourabi M, Nohair K, Traisnel M, Jama C, Bentiss F (2013) Electrochemical and XPS studies of the corrosion inhibition of carbon steel in hydrochloric acid pickling solutions by 3,5-bis(2-thienylmethyl)-4-amino-1,2,4-triazole. Corros Sci 75:123–133. https://doi.org/10.1016/j.corsci.2013.05.023
7. Lai C, Xie B, Zou L, Zheng X (2017) Adsorption and corrosion inhibition of mild steel in hydrochloric acid solution by S-allyl-O, O′-dialkyldithiophosphates. Results Phys 7:3434–3443. https://doi.org/10.1016/j.rinp.2017.09.012
8. Kumari P, Shetty P, Rao SA (2014) Corrosion inhibition effect of 4-hydroxy-N′-[(e)-(1h-indole-2-ylmethylidene)] benzohydrazide on mild steel in hydrochloric acid solution. Int J Corros. 2014:1–11. https://doi.org/10.1155/2014/256424
9. Khadom AA (2014) Effect of temperature on corrosion inhibition of copper - nickel alloy by tetraethylenepentamine under flow conditions. J Chil Chem Soc 59:2545–2549. https://doi.org/10.4067/S0717-97072014000300004
10. Gurudatt DM, Mohana KN, Tandon HC (2015) Adsorption and corrosion inhibition characteristics of some organic molecules containing methoxy phenyl moiety on mild steel in hydrochloric acid solution. Mater Discov 2:24–43. https://doi.org/10.1016/j.md.2016.03.005
11. Kini UA, Shetty P, Shetty D, Isloor MA (2011) Corrosion inhibition of 6061 aluminium alloy/SiCp composite in hydrochloric acid medium using 3-chloro-1-benzothiophene-2-carbohydrazide. Indian J Chem Technol 18:439–445
12. Abeng FE, Idim VD, Magu OE (2017) Adsorption and thermodynamic studies for corrosion inhibition of API 5L X-52 seel in 2 M HCl. World News Nat Sci 77:298–313
13. Udhayakala P (2014) Density functional theory calculations on corrosion inhibitory action of five azlactones on mild steel. J Chem Pharm Res 6:117–127
14. Lukovits I, Kálmán E, Zucchi F (2001) Corrosion inhibitors - correlation between electronic structure and efficiency. Corrosion 57:3–8. https://doi.org/10.5006/1.3290328
15. Rosliza R, Izman S (2011) SEM EDS characterization of natural products on corrosion inhibition of Al−Mg−Si alloy. Prot Met Phys Chem Surf 47:395–401. https://doi.org/10.1134/S2070205111030129
16. Hajjaji EF, Salim R, Messali M (2019) Electrochemical studies on new pyridazinium derivatives as corrosion inhibitors of carbon steel in acidic medium. J Bio- Tribo-Corr. 5:1–13. https://doi.org/10.1007/s40735-018-0195-3
17. Ouici HB, Benali O, Guendouzi A (2016) Experimental and quantum chemical studies on the corrosion inhibition effect of synthesized pyrazole derivatives on mild steel in hydrochloric acid. Res Chem Intermed 42:7085–7109. https://doi.org/10.1007/s11164-016-2520-0
18. Prathima S, Jyothi K, Rajitha S, Damodara N, Nithyananda A (2022) Investigation of the inhibition effect of newly synthesized pyrazoline derivative on mild steel in hydrochloric acid medium by experimental and theoretical approach. Chem Data Collect 10010. https://doi.org/10.1016/j.cdc.2021.100808
19. Ouici HB, Guendouzi A, Ferhat M, Benali O, Boudjellal F (2020) Corrosion inhibition of mild steel by newly synthesized pyrazole carboxamide derivatives in hcl acid medium: experimental and theoretical studies. J Electrochem Soc 167:155508. https://doi.org/10.1149/1945-7111/abab25
20. Arrouji SE, Alaoui KI, Zerrouki A (2016) The influence of some pyrazole derivatives on the corrosion behaviour of mild steel in 1M HCl solution. J Mater Environ Sci 7:299–309. ISSN: 2028-2508
21. Mechbal N, Belghiti ME, Benzbiria N (2021) Correlation between corrosion inhibition efficiency in sulfuric acid medium and the molecular structures of two newly eco-friendly pyrazole derivatives on iron oxide surface. J Mol Liq 331:115656. https://doi.org/10.1016/j.molliq.2021.115656

Corrosion Inhibition Studies on Mild Steel by Tetramethyl Ammonium Bromide in Hydrochloric Acid Medium

Vipul K. Sinha, Vanshika Ghai, Ayush K. Jha, B. Prayash Mohanty, Vishwanatha A. Devaranavadagi, Priti Gupta, Bijayani Panda, and I. C. Lekshmi

1 Introduction

Mild steel is a largely used material for various industrial applications. However, it has a tendency towards corrosion [1]. In several industrial processes, acids are used for the removal of rust and scales from the steel surfaces by pickling or descaling process [1]. Among different acids used for pickling, HCl and H_2SO_4 are the most widely used [2–4]. The advantage of using HCl is that it is comparatively economical and efficient to use [5]. The corrosion that occurs during pickling or descaling processes however causes high risk of structural damage along with high cost of repairing, inspection and replacement [4]. Using corrosion inhibitors is an adequate and cost-effective remedy here [5]. When coated on metal surfaces, the corrosion rate gets reduced through adsorbed inhibitors as they block active sites of corrosion. In the case of organic compound inhibitors, the adsorption properties of molecules determine the inhibition efficiency. The key role is performed by d-orbitals of the metal surface atoms and π-orbitals of the inhibitor interaction [6]. Quaternary ammonium salts showed significant inhibition efficiency when used as corrosion inhibitors, as reported in various studies [7, 8]. For mild steel, tetramethyl ammonium bromide (TMAB) in the concentration range of 10^{-7}–10^{-3} M was observed as excellent corrosion inhibitor particularly in presence of 1 N sulphuric acid containing solutions [9].

V. K. Sinha · V. Ghai · A. K. Jha · B. Prayash Mohanty · V. A. Devaranavadagi · B. Panda (✉)
Metallurgical Engineering, Department of Mechanical Engineering, CMR Institute of Technology, Bengaluru, India
e-mail: bijayani.p@cmrit.ac.in

V. K. Sinha · V. Ghai · A. K. Jha · B. Prayash Mohanty · P. Gupta · I. C. Lekshmi (✉)
Materials and Sensor, Department of Chemistry, CMR Institute of Technology, Bengaluru, India
e-mail: chaitanya.indira@plaksha.edu.in

I. C. Lekshmi
Plaksha University, Mohali, India

S. T. Aruna et al. (eds.), *Recent Advances in Electrochemical Science and Technology*, Springer Proceedings in Materials 47, https://doi.org/10.1007/978-981-97-3958-5_13

However, the inhibition effect of TMAB on mild steel in an HCl environment is not reported in the literature.

In the present work, an investigation of the corrosion inhibition action of TMAB on mild steel in 0.1 M HCl solution was conducted at room temperature (RT). Different techniques such as electrochemical impedance spectroscopy (EIS), potentiodynamic polarization (PDP), scanning electron microscopy (SEM) and Fourier transform infrared (FTIR) spectroscopy were used to record and analyze the corrosion results. Studies on the effect of different concentrations of HCl on mild steel corrosion were also conducted.

2 Experimental

2.1 Materials and Methods

For corrosion inhibition studies, mild steel sheets of thickness 5 mm were used. All reagents were used as procured with no additional purification and were of analytical grade. The chemicals used were HCl (AR grade 35.4%) solution, acetone and TMAB which was obtained from Spectrochem Pvt. Ltd. Mumbai, India. Double distilled water (DDW) received from the double distillation unit installed in the laboratory was used for preparing solutions.

An area of 30 mm $\times$ 30 mm of mild steel specimens was cut using Chop Saw (DEWALT, model D28870-B101), and their surfaces were ground using an angle grinder (BOSCH, model GWS 6-100). For a mirror finish, they were further polished successively using sandpaper (grit size 80), emery paper (grit size 80, 220, 400, 600, 800 and 1000) and a metallography polishing machine (Conation Technologies, model CPM-I), with intermittent brushing and cooling using water. Surface degreasing of samples was done using acetone followed by DDW on the polished surface and dried. They were finally stored under vacuum at ambient temperatures to avoid contamination of the polished surfaces. The polished and cleaned specimens were used to perform corrosion and control studies.

Figure 1 shows the structure of the corrosion inhibitor (TMAB) used in the present study. TMAB is a quaternary ammonium salt in its aqueous solutions and was readily soluble in HCl solution as observed during the experiment.

To prepare solutions of 0.1 M, 0.5 M and 1 M HCl, requisite amounts of AR grade 35.4% HCl were carefully added to DDW. Inhibitor (TMAB) solution concentrations used for the study were 0.01 M, 0.1 M and 0.2 M, prepared by adding requisite amounts of inhibitor salts weighed accurately (0.6162 g, 6.1620 g and 12.3240 g respectively) and dissolved in 400 ml of 0.1 M HCl solution prepared as above.

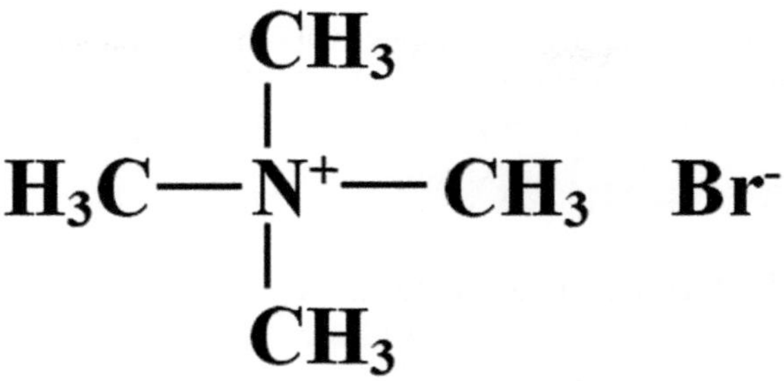

Fig. 1 Structure of TMAB

2.2 *Electrochemical Measurements*

For electrochemical measurements, mild steel specimens of area 30 mm × 30 mm were used as test electrodes after careful surface preparation as discussed in Sect. 2.1. To complete the three-electrode test circuit, a graphite rod was used as the counter electrode (CE), and a saturated calomel electrode (SCE) was used as the reference electrode (RE). At RT, experiments were performed in the freely aerated solution in the three-electrode test set-up. A surface area of 2.85 cm^2 of specimens used was exposed to the test solutions with or without the inhibitor. In the beginning, open circuit potential (OCP) was observed for 14,400 s (4 h) or till a stabilized potential was attained in each case. EIS measurements over the AC frequency range of 0.01 Hz to 100 kHz at the OCP using ±10 mV perturbation was then performed. As EIS is a non-destructive measurement technique, PDP experiments were run successively on the same specimens. The potential range used for PDP tests was −250 to + 1200 mV (versus OCP) with 1 mV/s scan rate. From PDP plots, corrosion potential (E_{corr}) and the corrosion current density (i_{corr}) were determined [4]. The process of determination of polarization resistance (R_p), and corrosion rate are discussed later in Sect. 3.2.

2.3 *Immersion Test and Corrosion Characterization*

The cut specimens of the same mild steel having 25 mm × 25 mm area were immersed in 0.1 M HCl solution in the absence and presence of 0.01 M, 0.1 M and 0.2 M TMAB solution for 15 days at RT, prepared by adding the requisite amounts of inhibitor salts weighed accurately (1.2324 g, 12.3240 g and 24.6480 g respectively) and dissolved in 800 ml of prepared 0.1 M HCl solution. The specimens were drawn out, completely washed with DDW and blow dried after the immersion period. Subsequently, they were dried in an air oven to remove moisture. For examination of the surface morphology of the specimens to find out the nature of the corrosion attack, SEM was used. Secondary Electrons (SE) detector was used for SEM imaging to study the surface topography [10]. The corrosion product collected from each specimen was filtered and washed using DDW separately. The dried corrosion products were later characterized by the FTIR technique.

3 Results and Discussion

3.1 EIS Measurements

Figure 2a shows the Nyquist plots obtained from EIS measurements on mild steel specimens in the absence and presence of varied TMAB concentrations at RT. Nyquist plots obtained from EIS data consisted of a semi-circular capacitive loop for all test solutions [11]. With the increase in concentration of TMAB added to 0.1 M HCl solution, there was a significant rise in the impedance value as evident from the increase in the semi-circle loop size.

At the interface of metal/solution, the double layer formed was dissimilar from that of a real capacitor due to irregularities present on the steel surface or leaky capacitance [2, 12]. Thus, the semi-circular capacitive loop deviated from an ideal semi-circle. A more accurate fit was obtained by replacing double-layer capacitance with a constant phase element (CPE), whose impedance could be expressed as [11, 12]:

$$Z_{\mathrm{CPE}} = \frac{1}{Y^{\circ}(j\omega)^{\alpha}} \tag{1}$$

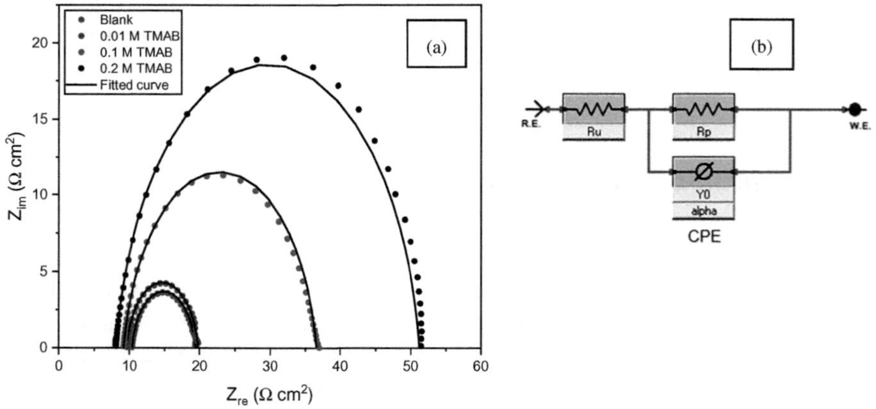

Fig. 2 **a** Nyquist plots of mild steel in 0.1 M HCl solution in the absence and presence of 0.01 M, 0.1 M and 0.2 M TMAB and **b** corresponding electrochemical circuit used to represent mild steel/solution interface in the absence and presence of the inhibitor

where Y° represented CPE constant, $j^2 = -1$, and ω and α denoted angular frequency and CPE component respectively that measured the extent of non-ideal behaviour of the capacitor from that of the ideal behaviour. For a pure capacitor, the value of α is 1.

The charge transfer of the double layer could be found using the following equation [12]:

$$C_{dl} = (Y^\circ {R_p}^{1-\alpha})^{\frac{1}{\alpha}} \quad (2)$$

where R_p denotes polarization resistance. The corresponding electrochemical circuit representing the mild steel/solution interface is shown in Fig. 2b. In the circuit, R_u corresponds to the electrolyte resistance, while R.E. and W.E. denote the reference electrode and working electrode, respectively. The results obtained from the EIS fittings for mild steel corrosion and inhibition studies are tabulated and shown in Table 1.

To analyze the inhibition efficiency of the process, the following equation was used [2]:

$$\eta\,\% = \frac{(R_p - R_p^\circ)}{R_p} \times 100 \quad (3)$$

where polarization resistance in the absence and presence of inhibitor was represented by $R_p{}^\circ$ and R_p, respectively.

In 0.1 M HCl test solution, it is evident from Table 1 that with the rise in concentration of TMAB, R_p values increased and CPE values decreased. The reduction in CPE values could be ascribed to the rise in electrical double layer thickness around the steel specimen surface or due to the decline in local dielectric constant [2]. This may be attributed to the adsorption of TMAB molecules at the interface of mild steel and the test solution [8, 12, 13]. This also implied that with the rise in TMAB concentration in 0.1 M HCl solution, inhibition efficiency rose as R_p increased. Thus, it was confirmed that in 0.1 M HCl solution, TMAB is an adequate inhibitor of mild steel. Similar results were reported elsewhere [4, 11], where specimens tested in 1.0 M HCl solution showed a rise in R_p value and decline in C_{dl} value with rise in concentration of inhibitors like Cefazolin (4.38×10^{-4} M, 6.57×10^{-4} M, 8.76×10^{-4} M) for mild

Table 1 Parameters obtained from EIS of mild steel in 0.1 M HCl solution in the absence and presence of 0.01 M, 0.1 M and 0.2 M TMAB

Concentration	R_p (Ω cm^2)	R_u (Ω cm^2)	Y° (S.s$^\alpha$)	α	C_{dl} (μF cm^{-2})	Inhibition efficiency (%)
Blank	25.79	29.58	0.0039	0.8732	829.9	–
0.01 M TMAB	29.13	27.29	0.0027	0.8882	603.9	11.47
0.1 M TMAB	78.26	26.12	0.0013	0.8887	308.3	67.05
0.2 M TMAB	123.03	23.01	0.0007	0.9054	175.4	79.04

steel and di-quaternary ammonium salts (1×10^{-5}–5×10^{-3} M). The salts acted as mixed-type inhibitors in 1 M HCl solution due to a rise in inhibition efficiency in each of these studies.

Figure 3 shows the Nyquist plot of mild steel at varied concentrations of HCl in the absence of the inhibitor obtained from EIS experiment at ambient temperature. For mild steel, the experimental results attained by EIS in 0.1 M, 0.5 M and 1 M HCl solution without TMAB are shown in Table 2. It was found that polarization resistance (R_p) decreased and C_{dl} value enhanced with the rise in the concentration of HCl solution concurrent to higher current flow at higher concentrations of HCl (corrosion medium). The electrolyte resistance (R_u) also decreased considerably with the rise in the concentration of HCl, indicated by the shifting of the Nyquist plot towards the left.

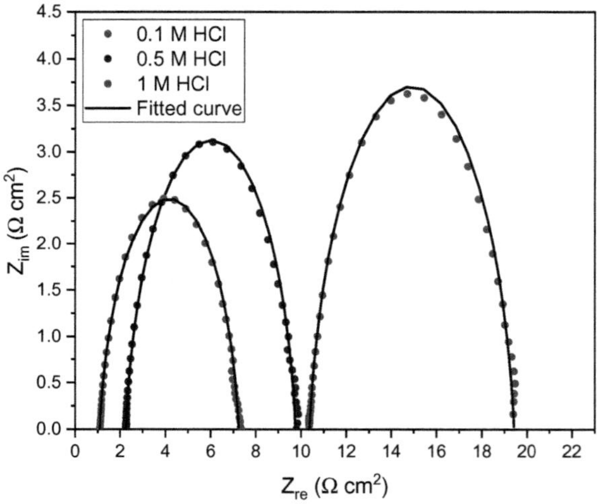

Fig. 3 Nyquist plot of mild steel in 0.1 M, 0.5 M and 1 M HCl solution without TMAB

Table 2 Parameters obtained from EIS of mild steel in 0.1 M, 0.5 M and 1 M HCl solution without TMAB

Concentration	R_p (Ω cm^2)	R_u (Ω cm^2)	$Y°$ ($S.s^{\alpha}$)	α	C_{dl}(μF cm^{-2})
0.1 M HCl	25.79	29.58	0.0039	0.8732	829.9
0.5 M HCl	21.39	6.42	0.0041	0.8843	920.1
1 M HCl	17.50	3.14	0.0050	0.8672	1032.0

3.2 Potentiodynamic Polarization Measurements

Figure 4 shows the PDP curves for mild steel in the absence and presence of varied TMAB concentrations at RT. The electrochemical reactions in an acidic medium for mild steel are as follows [12]:

$$\text{At Anode: } Fe \leftrightarrow Fe^{2+} + 2e^{-} \tag{4}$$

$$\text{At Cathode: } 2H^{+} + 2e^{-} \leftrightarrow H_2 \tag{5}$$

On linear portions of the anodic and cathodic regions of the curve, tangents were drawn using EC-Lab software. The intersection point of the extrapolated tangents gives E_{corr} on the Y-axis and i_{corr} on the X-axis as shown in the inset image in Fig. 4 [14]. Using the slopes of the anodic region (β_a) and cathodic region (β_c), polarization resistance was determined, as shown in the equation below [15]:

$$R_p = \frac{\beta_a \times \beta_c}{2.303 \times i_{\text{corr}} \times (\beta_a + \beta_c)} \tag{6}$$

Using the following equation, the corrosion rates were consequently obtained [16]:

$$\text{Corrosion rate (mpy)} = \frac{K \times i_{\text{corr}} \times EW}{\rho} \tag{7}$$

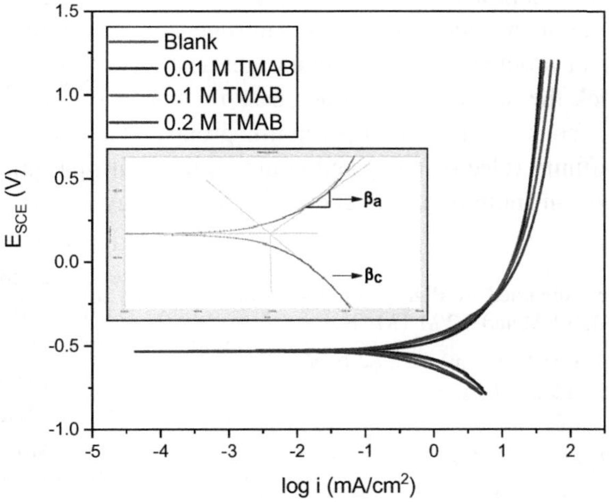

Fig. 4 PDP of mild steel in 0.1 M HCl solution in the absence and presence of 0.01 M, 0.1 M and 0.2 M TMAB. The Inset image shows PDP of mild steel in 0.1 M HCl solution in the presence of 0.1 M TMAB

where ρ denoted the density of the mild steel, $K = 0.1288$ mpy g/μA cm and EW denoted the equivalent weight of the mild steel.

According to studies mentioned in Refs. [2, 12], if the variation in E_{corr} is above 85 mV in comparison to the blank solution, then the inhibitor can be feasibly categorized as an anodic or cathodic type inhibitor. In this study, the shift in E_{corr} was 10 mV after adding TMAB in 0.1 M HCl solution which implied that TMAB was a mixed type of inhibitor. As a mixed type of inhibitor, the hydrogen evolution reaction at the cathode surface as well as the anodic iron dissolution reactions related to mild steel corrosion was controlled by TMAB [9]. It can be seen that TMAB affected both the Tafel slopes and suppressed anodic and cathodic reactions [12] associated with the corrosion of mild steel specimens in 0.1 M HCl solution. The results obtained from PDP are tabulated in Table 3. In potentiodynamic polarization experiments, the following equation was used to deduce the inhibition efficiency [2]:

$$\eta\,\% = \frac{(i^{\circ}{}_{corr} - i_{corr})}{(i^{\circ}{}_{corr})} \times 100 \tag{8}$$

where corrosion current density in the absence and presence of inhibitor was denoted by $i^{\circ}{}_{corr}$ and i_{corr}.

As shown in Table 3, with the rise in the concentration of TMAB in 0.1 M HCl solution, i_{corr} value reduced indicating the rise in the inhibition efficiency. The percentage inhibition efficiencies in the presence of 0.01 M, 0.1 M and 0.2 M TMAB were 5.72, 61.23 and 82.94, respectively. The inhibition could be designed to occur on the mild steel surface through the adsorption of inhibitor molecules that form a protective layer [2]. The extent of molecular adsorption and the inhibitor efficiency depends on several factors like the shape and size of the molecules, the component atoms and their orientation and the electric charges present on them [4]. Quaternary ammonium salts contain heteroatoms such as nitrogen, oxygen, sulfur or phosphorus, multiple bonds, or aromatic rings that could get adsorbed on the metal surface in contact and block the active sites and decrease the corrosion rate. As surfactants, the structure also consists of a polar hydrophilic group and a non-polar hydrophobic group, whose affinity (electrostatic) and compatibility with the mild steel surface decide the corrosion inhibition efficiency. The free energy change of transferring

Table 3 Parameters obtained by PDP of mild steel in 0.1 M HCl solution in the absence and presence of 0.01 M, 0.1 M and 0.2 M TMAB

Concentration	β_a (mV/decade)	β_c (mV/decade)	E_{corr} (mV)	i_{corr} (μA/cm^2)	R_p (kΩ cm^2)	Corrosion rate (mpy)	Inhibition efficiency (%)
Blank	277.9	273.9	−531.8	839.6	0.071	389.2	–
0.01 M TMAB	239.5	240.8	−526.7	791.6	0.066	366.9	5.72
0.1 M TMAB	153.9	184.5	−527.6	325.5	0.112	150.9	61.23
0.2 M TMAB	132.5	109.0	−531.3	143.2	0.181	66.4	82.94

a hydrocarbon chain from water to the solid surface also becomes decisive. The molecular structure of TMAB molecule consists of TMA^+ as the cation and Br^- as the anion. The mild steel surface in the presence of HCl solution is positively charged [12], that could attract the negatively charged bromide ions from the TMAB to adsorb onto the surface converting the steel surface to be more negative [7–11]. Adsorption of TMAB cations then occurred on the now negatively charged steel surface due to coulombic attraction [7–11]. Thus, the adsorption of inhibitor could be facilitated through the interaction between the transition metal d-orbitals in the mild steel and π-orbitals of the inhibitor molecules forming a protective layer and decelerating the corrosion reaction. The scope of corrosion inhibition from TMAB on mild steel surface at higher concentrations of HCl needs to be investigated further.

On comparing the PDP results of mild steel in the present study with those from the literature [11, 12] at 5 M and 1 M HCl solution, respectively, it was observed that i_{corr} value declined with a rise in inhibitor concentration. The substituted phenyltetrazole from the literature at 10^{-4}–10^{-3} M concentration acted as cathodic type inhibitor [12], while di-quaternary ammonium salts in the concentration range 1×10^{-5}–5×10^{-3} M worked as mixed-type inhibitors [11]. The range of inhibition efficiencies obtained was 67.4–91.2% for mild steel in the former case and 41.5–97.2% in the latter case to conclude as effective corrosion inhibitors. The inhibition efficiency exhibited by TMAB in our study at 0.1 M and 0.2 M were comparable to the above literature.

Figure 5 shows the PDP curves for mild steel at different concentrations of HCl solution and without inhibitor at ambient temperatures. With the rise in the concentration of HCl in the solution, i_{corr} was seen to rise. This is due to the rise in active species in the acid solution accelerating the corrosion reaction [17].

The results obtained from PDP are tabulated in Table 4. The variation in the corrosion rate of mild steel exposed to 0.1 M HCl solution as expressed in mils per year (mpy) units and with respect to varying concentrations of TMAB is exhibited

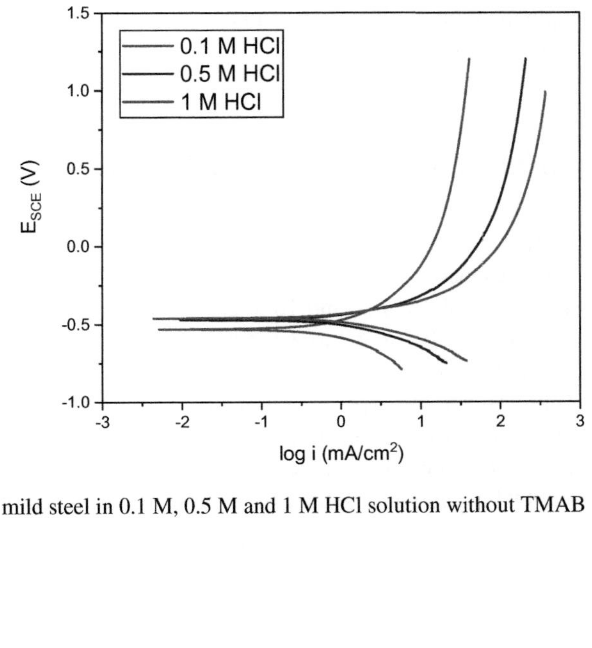

Fig. 5 PDP of mild steel in 0.1 M, 0.5 M and 1 M HCl solution without TMAB

Table 4 Parameters obtained by PDP of mild steel in varied concentrations of HCl without TMAB

Concentration	β_a (mV/decade)	β_c (mV/decade)	E_{corr} (mV)	i_{corr} (μA/cm^2)	R_p (kΩ cm^2)	Corrosion rate (mpy)
0.1 M HCl	277.9	273.9	−531.8	839.6	0.071	389.2
0.5 M HCl	173.3	206.4	−473.8	1386.3	0.030	642.6
1 M HCl	160.1	197.2	−461.3	1971.6	0.019	913.9

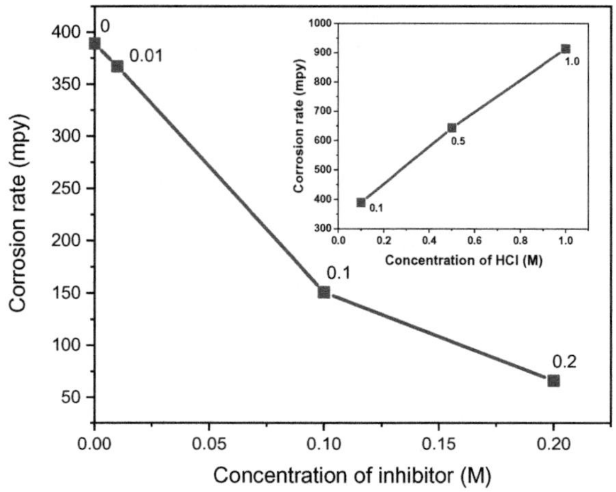

Fig. 6 Graph showing the variation of corrosion rate of mild steel in mils per year (mpy) with respect to the concentration of TMAB inhibitor in 0.1 M HCl. The Inset image shows the variation of corrosion rate of the same sample with respect to the concentration of HCl solution

in Fig. 6. The curve displayed a decreasing trend in the corrosion rate of mild steel in 0.1 M HCl solution on addition of TMAB due to their inhibiting effect in agreement with other results discussed in this study. The inset graph showed the expected linear variation of the corrosion rate of the mild steel with HCl concentration. As explained previously, this happened due to the increase in active species in the acidic solution that accelerated the corrosion reaction [17].

3.3 Fourier Infrared Spectroscopy (FTIR) Measurements

FTIR plots of the rust collected from mild steel specimens tested in 0.1 M HCl solution and in the presence of 0.01 M, 0.1 M and 0.2 M TMAB are shown in Fig. 7. FTIR absorption peaks at 749 and 1020 cm^{-1} were indicative of γ-FeOOH, while the peak at 470 cm^{-1} belonged to δ-FeOOH highlighting the rusting of the steel sample that occurred on immersing in 0.1 M hydrochloric acid [18]. The blank sample tested in 0.1 M HCl had sharper absorption peaks for γ-FeOOH due to higher corrosion on the specimen surface. On adding TMAB the peak intensity of γ-FeOOH decreased based on the concentrations suggesting the inhibition of the rusting process. Interestingly, γ-FeOOH is known to be the active phase of the corrosion product and

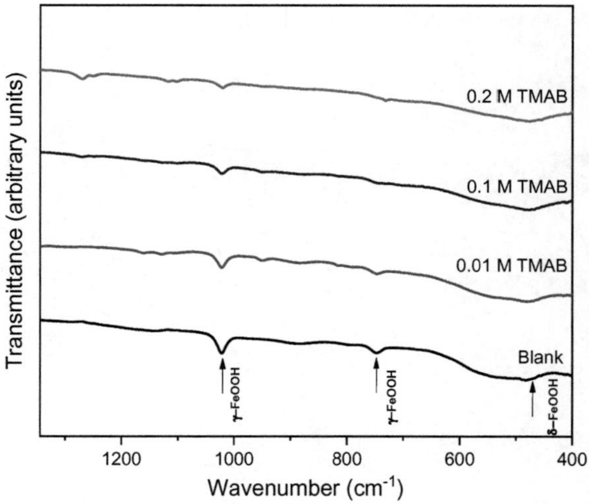

Fig. 7 FTIR spectra of rust formed on mild steel in 0.1 M HCl in the absence and presence of 0.01 M, 0.1 M and 0.2 M TMAB

a higher amount of the same in the corrosion product indicated lower degree of the sample protection. δ-FeOOH phase, on the other hand, is more amorphous and hence protective in nature. This phase was found to have similar intensity in all rust samples tested.

3.4 SEM Analysis

SEM images of mild steel surface after immersion in 0.1 M HCl solution for 15 days at RT in (a) absence and in presence of (b) 0.01 M, (c) 0.1 M and (d) 0.2 M TMAB are shown in Fig. 8. In the absence of an inhibitor, the mild steel was severely damaged by the attack of acid solution with an etched appearance as could be seen from Fig. 8a, revealing lamellar pearlitic microstructure. In the presence of TMAB however, the surface appeared to be relatively smooth as shown in Fig. 8b–d. As the concentration of TMAB was enhanced, the damage due to corrosion of steel was reduced. The inhibition was thought to have occurred because of the adsorption of the inhibitor molecules onto the mild steel surface that suppressed the cathodic and anodic reactions of the corrosion process [1].

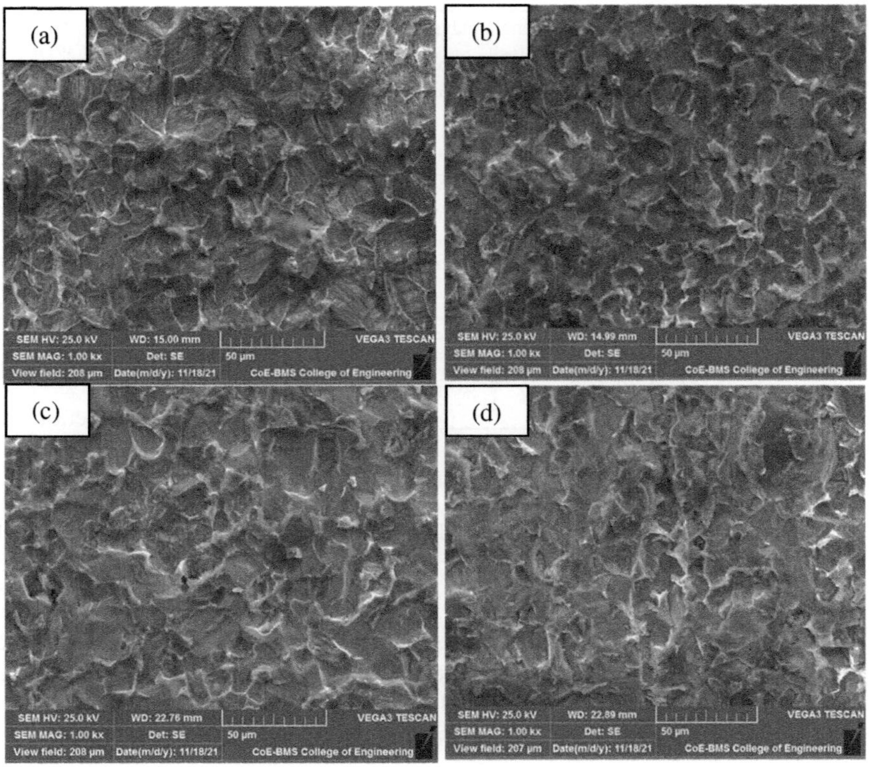

Fig. 8 SEM images of mild steel surface after immersion in 0.1 M HCl solution for 15 days in **a** absence and in presence of **b** 0.01 M, **c** 0.1 M and **d** 0.2 M TMAB inhibitor

The attack was least when TMAB concentration in acid solution reached 0.2 M. This was also ascertained by composition analysis of the corroded steel specimen carried out by energy dispersive X-ray (EDX) analyser that confirmed higher amount of Fe on the corroded mild steel surface after acid exposure containing 0.2 M TMAB (see Table 5). Similar trends were discussed by Khamis et al. [7] and Dubey et al. [9] respectively for mild steel exposed to 0.5 M H_2SO_4 for 30 min and 1 N H_2SO_4 for 24 h respectively. In the absence of inhibitors, the steel surface appeared severely damaged, while the presence of inhibitors such as CTAB (7×10^{-6}–1×10^{-3} M) and TMAB (10^{-7}–10^{-3} M) reduced the corrosion drastically making the surface relatively smooth.

Table 5 Semi-quantitative compositional analysis of corroded steel specimen surface; carried out by energy dispersive X-ray (EDX) analyser

Element	Composition (wt.%)						
	C	O	Si	P	Cl	Mn	Fe
0.1 M HCl	–[a]	20.83	0.27	0.07	0.16	1.29	60.43
0.1 M HCl+ 0.2 M TMAB	–[a]	5.61	0.30	0.23	0.20	1.52	74.00

[a] Carbon cannot be accurately determined by EDX analysis

Mild steel sheets used in the study were cold-rolled. The attack along the rolling direction was expected to be higher due to the high amount of stress present along this direction. SEM studies were therefore carried out along the rolling direction and the intensity of localized corrosive attack was compared. In the blank test, a deep localized attack was observed as the steel specimen was directly exposed to 0.1 M HCl (see Fig. 9a). The corresponding magnified SEM image shows the depth of attack on the steel surface. In Fig. 9b–d, the SEM images of the acid attack along the rolling direction in the presence of TMAB are shown. The white arrows represented the corrosion while the depth of attack could be understood from the corresponding magnified images. The extent of the localized attack was substantially reduced and was seen to be shallow as can be seen from the comparing figures when inhibitor (TMAB) was added to the acid solution [7, 9].

4 Conclusions

TMAB is a promising inhibitor for mild steel. 0.01 M, 0.1 M and 0.2 M concentrations of TMAB in 0.1 M HCl solution were found to provide significant inhibition efficiency. The inhibition was enhanced with the rise in the concentration of the bromide salt molecules added that were adsorbed on the mild steel surface exposed to acid solution and formed a protective layer. Potentiodynamic polarization results displayed that in the presence of 0.1 M HCl solution, TMAB molecules performed as mixed-type inhibitors. Further, the EIS measurements of the specimens indicated a significant increase in the impedance on the addition of TMAB. SEM and EDAX studies of the corroded samples tested for varied concentrations of TMAB in 0.1 M HCl also ascertained substantial reduction in general corrosion as well as in localized corrosion attacks around the rolling direction of the sheet. This was corroborated by reduced rust formation. Investigations were also conducted to find the effect of varied HCl solution concentrations on mild steel corrosion. The corrosion rate increased linearly with the acid solution concentration that occurred due to a rise in active species in the acid that accelerated the corrosion reaction.

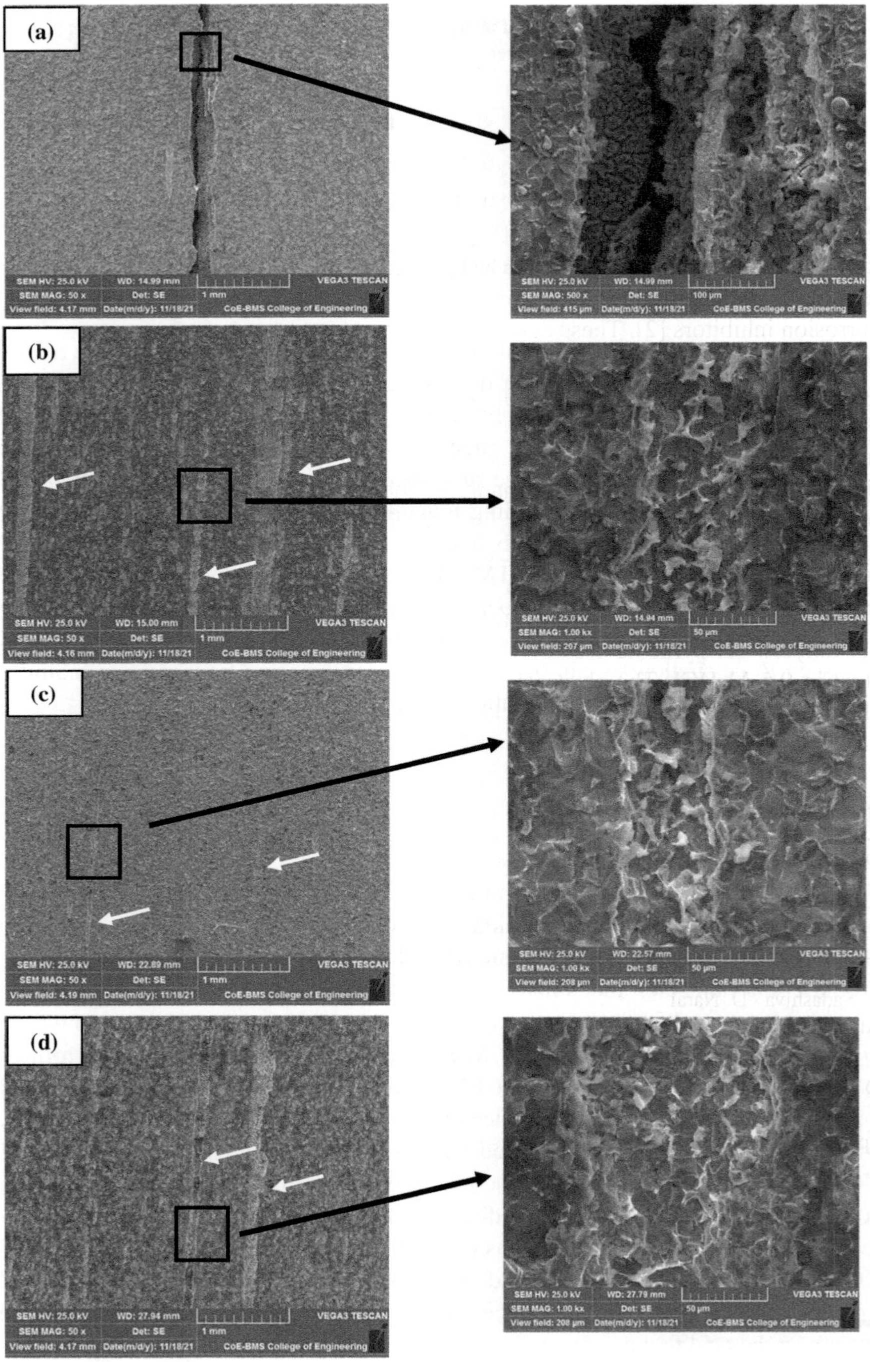

Fig. 9 Images attained by SEM of mild steel surface after immersion in 0.1 M HCl solution for 15 days in **a** the absence of TMAB and in presence of **b** 0.01 M, **c** 0.1 M and **d** 0.2 M TMAB showing localized attack along the rolling direction

Acknowledgements The authors are greatly thankful to DST TMD (Grant No. DST/TMD(EWO)/OWUIS-2018/RS-22(G)) and NRB (Grant No. MAT/450/19-20) for providing support and facilities to conduct the experiments.

References

1. Bentiss F, Traisnel M, Lagrenee M (2000) The substituted 1,3,4-oxadiazoles: a new class of corrosion inhibitors of mild steel in acidic media. Corros Sci 42:127–146. https://doi.org/10.1016/S0010-938X(99)00049-9
2. Döner A, Kardas G (2011) N-Aminorhodanine as an effective corrosion inhibitor for mild steel in 0.5 M H_2SO_4. Corros Sci 53:4223–4232. https://doi.org/10.1016/j.corsci.2011.08.032
3. Ramananda Singh M, Bhrara K, Singh G (2008) The inhibitory effect of diethanolamine on corrosion of mild steel in 0.5 M sulphuric acidic medium. Port Electrochim Acta 26(6):479–492. https://doi.org/10.4152/pea.200806479
4. Singh AK, Quraishi MA (2010) Effect of Cefazolin on the corrosion of mild steel in HCl solution. Corros Sci 52:152–160. https://doi.org/10.1016/j.corsci.2009.08.05
5. Singh DDN, Singh TB, Gaur B (1995) The role of metal cations in improving the inhibitive performance of hexamine on the corrosion of steel in hydrochloric acid solution. Corros Sci 37:1005–1019. https://doi.org/10.1016/0010-938X(95)00010-H
6. Quraishi MA, Sharma HK (2002) 4-Amino-3-butyl-5-mercapto-1,2,4-triazole: a new corrosion inhibitor for mild steel in sulphuric acid. Mater Chem Phys 78(1):18–21. https://doi.org/10.1016/S0254-0584(02)00313-9
7. Khamis A, Mahmoud Saleh M, Mohamed Awad I, El-Anadouli BE (2013) Inhibitory action of quaternary ammonium bromide on mild steel and synergistic effect with other halide ions in 0.5 M H_2SO_4. J Adv Res 5:637–646. https://doi.org/10.1016/j.jare.2013.09.003
8. Prathibha BS, Kotteeswaran P, Bheema Raju V (2013) Study on the inhibition of mild steel corrosion by quaternary ammonium compound in H_2SO_4 medium. Res J Recent Sci 2(4):1–10
9. Dubey AK, Singh G, Tiwari K (2012) Corrosion inhibition of mild steel in sulphuric acid solutions by using tetra methyl ammonium bromide (TMAB). J Chem Eng Mater Sci 3(5):86–95. https://doi.org/10.5897/JCEMS11.032
10. Kejzlar P, Švec M, Macajová E (2014) The usage of backscattered electrons in scanning electron microscopy. Manuf Technol 14(3):333–336. https://doi.org/10.21062/ujep/x.2014/a/1213-2489/MT/14/3/333
11. Hegazy MA, Abdallah M, Awad MK, Rezk M (2014) Three novel di-quaternary ammonium salts as corrosion inhibitors for API X65 steel pipeline in acidic solution. Part I: Experimental results. Corros Sci 81:54–64. https://doi.org/10.1016/j.corsci.2013.12.010
12. El Kacimi Y, Azaroual M A, Touir R, Galai M, Alaoui K, Sfaira M, EbnTouhami M, Kaya S (2017) Corrosion inhibition studies for mild steel in 5.0 M HCl by substituted phenyltetrazole. Euro-Mediterr J Environ Integr 2:1. https://doi.org/10.1007/s41207-016-0011-8
13. Al-Amiery AA, Abdul AH, Kadhum AH, Alobaidy M, Mohamad AB, Hoon PS (2014) Novel corrosion inhibitor for mild steel in HCl. Materials 7:662–672. https://doi.org/10.3390/ma7020662
14. Elhebshi A, El Nemr A, Mohamed El-Deab S, Ashour I (2019) CBG-HClas a green corrosion inhibitor for low carbon steel in 0.5 M H_2SO_4 with and without 0.1 M NaCl. Desal Water Treat 164:240–248. https://doi.org/10.5004/dwt.2019.24446
15. Vishwanatha AD, Shivanna DM, Panda B (2021) Corrosion behaviour of in situ AlxNiy reinforced AA6061 composite. Key Eng Mater 882:96–103. https://doi.org/10.4028/www.scientific.net/KEM.882.96
16. Panda B, Balasubramaniam R, Dwivedi G, Mahapatra S (2008) Corrosion of novel rail steels in 3.5 % NaCl solution. Trans Indian Inst Met 61(2):1. https://doi.org/10.1007/S12666-008-0005-6

17. Orubite- Okorosaye K, Oforka NC (2004) Corrosion inhibition of zinc on HCl using nypa fruticans wurmb extract and 1,5 diphenyl carbazone. J Appl Sci Environ Mgt 8(1):57–61. https://doi.org/10.4314/jasem.v8i1.17228
18. Panda B, Balasubramaniam R, Dwivedi G (2008) On the corrosion behaviour of novel high carbon rail steels in simulated cyclic wet-dry salt fog conditions. Corros Sci 50:1684–1692. https://doi.org/10.1016/j.corsci.2008.02.021

Cost-Effective Synthesis of Zinc Oxide Nanorods and Their Properties

Pratibha S. Kanavi, Sunil Meti, R. H. Fattepur, and Veerabhadragouda B. Patil

1 Introduction

Presently, in electronic industries, one-dimensional (1D) semiconductor nanostructures such as rods, wires, belts, and tubes are explored as novel materials with multifunctionality. These are stimulating the research in the design of tailored materials for various potential applications [1]. Nanorods can be synthesized using metals/non-metals/compounds and the development of these kinds of materials is much easier as compared to other structures [2]. Nowadays, researchers are focusing on the size of materials, and they are bringing novel materials for future applications by achieving the size of materials to the nanoscale. Considerably these nanoscale materials possess unique features than the material at bulk scale [2]. Fabrication of electronic nanodevices is quite difficult. Since the last decade, plenty of electronic devices have been out to reach considerable applications [3, 4]. The vapor-solid and vapor-liquid-solid methods are most widely used to synthesize semiconducting materials of group II–VI, III–V, and doped hetero junctions [4–7]. In addition, other techniques like template-induced growth, catalytic growth by laser ablation, carbon-based diluters solution growth [8, 9] are used to achieve 1D nanostructures of different materials.

P. S. Kanavi (✉)
C. N. R. Rao Research Centre, Basaveshwar Science College, Bagalkot, Karnataka 587101, India
e-mail: pratibharamgeri@gmail.com

S. Meti
Department of Centre for Nano Science and Engineering (CeNSE), Indian Institute of Science, Bengaluru 560012, India

R. H. Fattepur
Raichur University, Raichur, Karnataka 584101, India

V. B. Patil
Institute of Energetic Materials, Faculty of Chemical Technology, University of Pardubice, 53210 Pardubice, Czech Republic
e-mail: iamveerabhadraa@gmail.com

S. T. Aruna et al. (eds.), *Recent Advances in Electrochemical Science and Technology*, Springer Proceedings in Materials 47, https://doi.org/10.1007/978-981-97-3958-5_14

ZnO is an applied semiconductor material and exhibits a band gap of 3.36 eV. ZnO is appropriate for transparent and optoelectronic-related technology applications such as a transparent conducting oxide in varistors, gas sensing, and catalysis activity [10–14].

Recently, ZnO nanorods possessing greater binding energy (60 meV) and a wide band gap (3.36 eV) have been prepared. ZnO received considerable research attention due to its unique features like UV emission, piezoelectric response, and electrical conductivity. Rather than these applications, it also has considerable application in the medical field. The ZnO is a good semiconducting material that is mostly used to fabricate light-emitting diodes, biosensors, transparent electrodes, UV filters, etc. [15]. The various research investigations reported that the surface morphology of 1D ZnO ordered materials enhances the thermo-optical properties and hence 1D ZnO seems the most advanced material among other dimension scales [16, 17]. Several types of 1D ZnO nanostructures such as nanobelts, nanodots, nanotubes, nanorods, and nanowires have been investigated [18, 19]. ZnO nanorods act as building blocks for futuristic applications in electronics and photonics [20]. The electrical and mechanical properties of the nanomaterials can be tuned by improving the dimensionality and quantum confinement of material structures [1]. This article deals with the construction of 1D ZnO nanorods through cost-effective simple co-precipitation method. These nanorods were confirmed by different characterization techniques. Structural and morphological properties have been studied.

2 Materials and Methods

2.1 Materials

All the chemicals were of analytic grade and procured from Sigma-Aldrich, Bangalore, India. All the solutions were prepared with distilled water (DI). For the preparation of ZnO nanorods, zinc nitrate ($Zn(NO_3)_2{\cdot}6H_2O$) and sodium hydroxide (NaOH); were utilized without further purification.

2.2 Synthesis of ZnO Nanorods

The ZnO was synthesized via an inexpensive and simple co-precipitation method. In a typical procedure, 0.1 m of NaOH was used as the precipitating agent and was slowly included dropwise into $Zn(NO_3)_2{\cdot}6H_2O$ (0.1 M) precursor dissolved in 10 ml of distilled water maintaining the temperature at 70 °C under constant and vigorous magnetic stirring up to 1 h until a white precipitate was developed. Further, at 1000 rpm for 10 cycles, the resultant product was centrifuged for 3 min and filtered and washed with ethanol several times [18]. The final product was kept for drying at 150 °C for 3 h to get the powder sample.

2.3 Experimental Methods

The phase purity, and the crystallite size of the synthesized ZnO nanorods were analyzed using X-ray diffractometry (XRD) with CuK_α ($\lambda = 1.54$ Å) source. The surface morphology study of ZnO nanorods was carried out by High-resolution transmission electron microscopy (HRTEM) and energy dispersive spectroscopy (EDS) analysis. The Fourier transform infrared (FTIR, model PERKIN ELMER) spectrometer was used to obtain the FTIR spectrum of the sample. UV-visible spectrophotometer Model-2201 was used to record the UV-visible spectrum in the wavelength range of 200–800 nm.

3 Results and Discussions

3.1 XRD Analysis of ZnO

The XRD analysis of ZnO nanorods shown in Fig. 1. The main peaks observed at $2\theta = 31.80, 33.90, 36.24, 47.55, 56.60, 62.90$ and $67.94°$ are corresponding with the planes (100), (002), (101), (102), (110), (103) and (112), respectively [21]. No additional peaks are observed in Fig. 1, which confirms that ZnO nanorods formed with crystalline hexagonal wurtzite structure with lattice constants, a = b = 3.249 Å and c = 5.209 Å (JCPDS card no, 36-1451). The average crystalline size of ZnO nanorods was evaluated to be about 29 nm using the Debye Scherrer formula (Eq. (1)) [22].

$$D \approx \frac{0.9\lambda}{\beta_D \cos\theta} \tag{1}$$

where D—Average crystallite size, λ—Wavelength of X-ray used (1.50406×10^{-10} m), β_D—The full width at half maximum of the most intense reflection peak in radian and θ—Bragg's angle in radian.

3.2 EDS Analysis

Figure 2 illustrates the chemical composition of the synthesized ZnO nanorods. From the EDS spectrum, it is observed that the elemental compositions were indeed as Zn

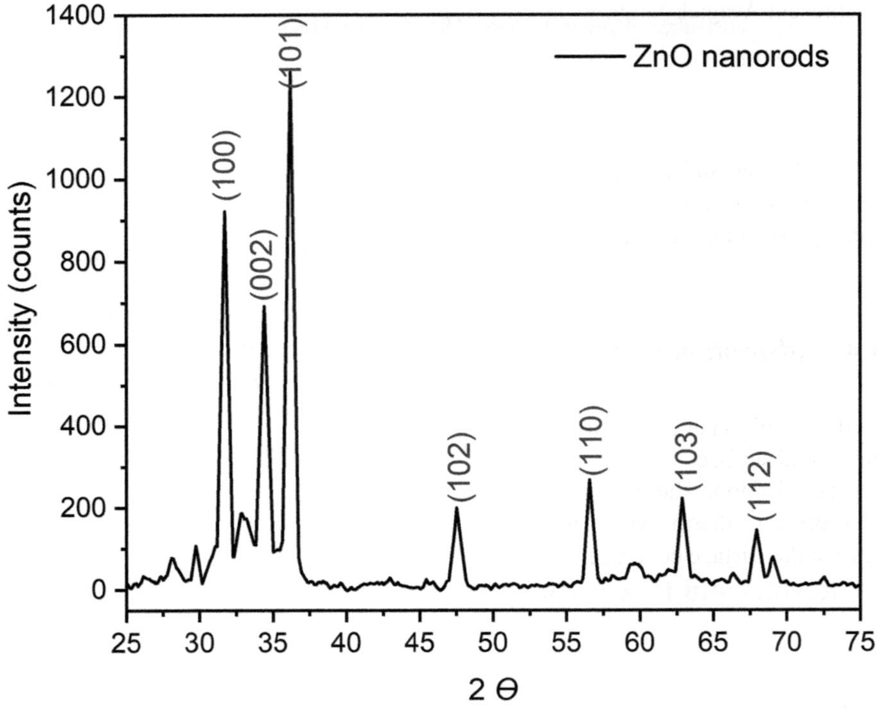

Fig. 1 XRD analysis of ZnO nanorods

and O and also the small quantities of additional elements like C, Ca, Si and Cu were also detected (Fig. 2), which may be due to the organic leftover components, whereas the C element correlates with the carbon-coated grid. The EDS results of ZnO nanorods composition revealed that ZnO was entrapped within the cured epoxy network.

3.3 Morphological Analysis

A typical HRTEM analysis was used to evaluate the microstructure analysis of ZnO nanorods, which clearly shows the morphological structure of the ZnO nanorods (Fig. 3a–b) synthesized by the cost-effective co-precipitation method. It is evident from the microscopic images that the synthesized ZnO nanorods are of uniform length and width and it discloses the confined boundaries of the ZnO nanorods. The selected area diffraction (SAED) pattern of uniform ZnO nanorods (NR's) is shown in Fig. 3c which explores the crystalline nature of nanorods with the lattice planes (100), (110) and (112) and this information is well in agreement with the XRD (Fig. 1) analysis showing the ordered structure of ZnO [23].

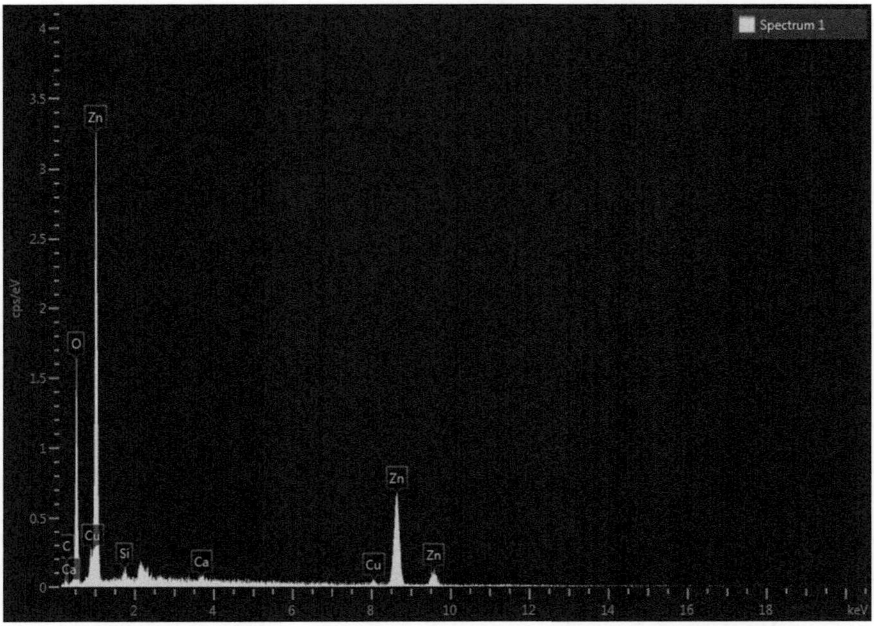

Fig. 2 EDS spectrum of ZnO

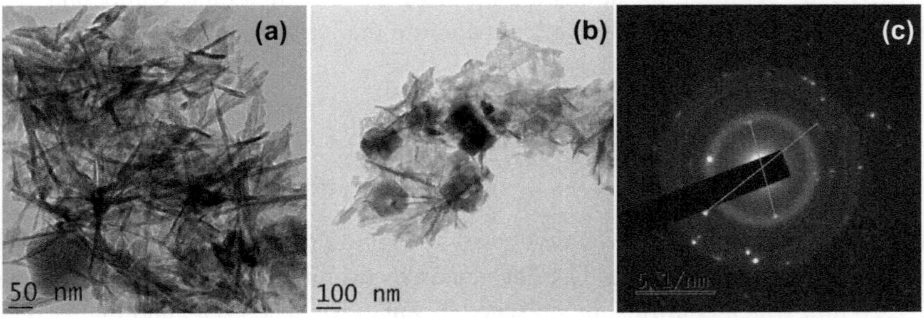

Fig. 3 **a**, **b** HRTEM analysis with different magnifications and **c** SAED pattern of the synthesized zinc oxide

3.4 FTIR Spectrum of ZnO nanorods

Synthesized ZnO nanorods were analyzed by FTIR in the range of 400–4000 cm^{-1} as shown in Fig. 4. The band at 3457 cm^{-1} is correlated with the O–H stretching existence. While the peaks at 2923 indicate C–H stretching and 2853 cm^{-1} is attributed to C–H bending vibration, respectively. The bands at 1383 and 1322 cm^{-1} correspond to C=O asymmetric vibration, and the band at 1624 cm^{-1} is responsible for the symmetric C=O stretching vibration. The adsorption peak corresponding to tetrahedral coordination of Zn–O bond seen at 875 cm^{-1}. The major bonds at 608

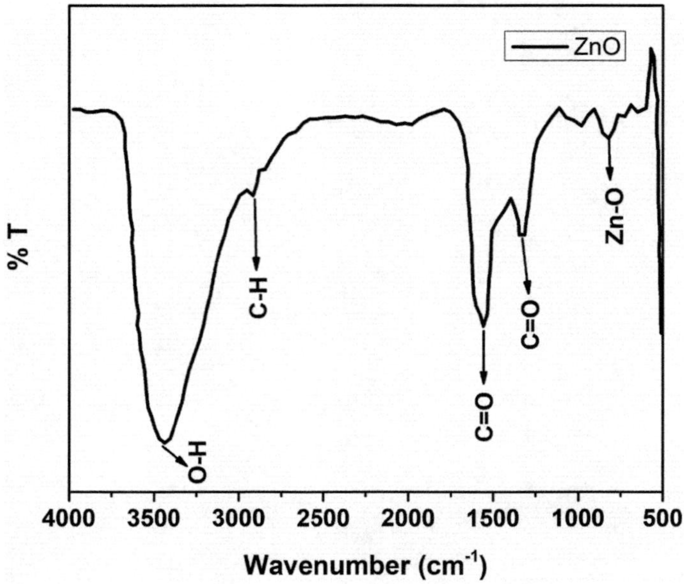

Fig. 4 FTIR spectra of ZnO nanorods

and 731 cm^{-1} indicate the stretching mode of the ZnO inorganic compound. The functional groups in the FTIR spectrum confirm the formation of ZnO [24, 25].

3.5 UV-Visible Spectra of ZnO nanorods

Figure 5 illustrates the UV-Visible optical absorption spectra of the ZnO nanorods recorded at room temperature. The ZnO nanorods exhibited a strong absorption band at ~220 nm, which confirms the presence of ZnO [26]. The broad and intense UV-absorption peak at 220 nm corresponds to the characteristic peak of the hexagonal wurtzite structure of ZnO [14]. This can lead to the development of quantum confinement of excitons present in the sample compared with bulk ZnO particles. The result reveals that the simple co-precipitation method exhibits stable ZnO nanorods [26, 27]. Using the Tauc's plot, the absorption band gap of ZnO was estimated (Fig. 6). The evaluated band gap of ZnO was determined to be 2.68 eV, which enhances its various applications such as gas sensors, batteries, capacitors, light-emitting diodes, electrical conductivity, biosensors, etc.

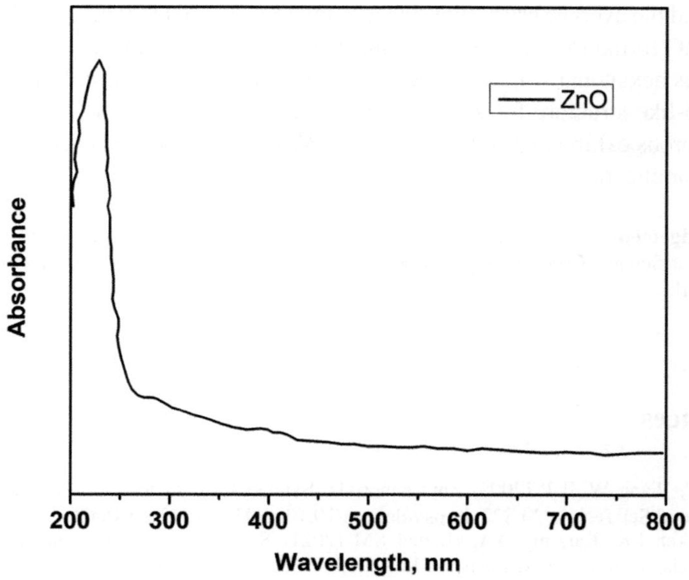

Fig. 5 UV-visible spectrum of ZnO nanorods

Fig. 6 Tauc's plot to evaluate the band gap of ZnO nanorods

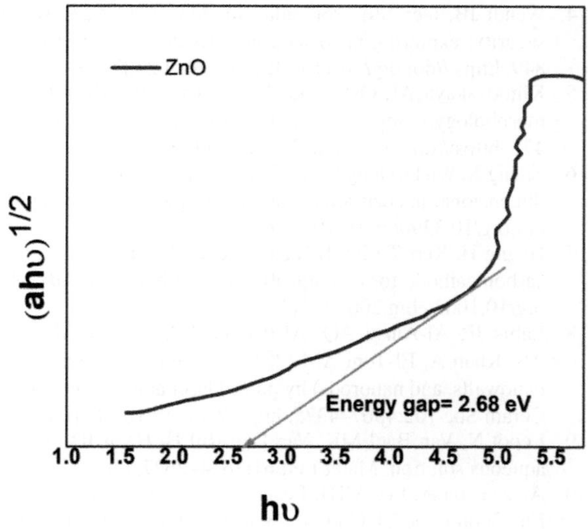

4 Conclusions

ZnO nanorods with uniform length were successfully prepared via a simple and single-step co-precipitation method and thoroughly characterized. Showed interesting characteristics; XRD analysis confirmed the good crystallinity of the ZnO sample and the average crystallite size was found to be about 29 nm, FTIR spectrum shown ZnO formation and UV spectrum strong absorption band at ~220 nm further it indicates hexagonal wurtzite structure of ZnO. Morphology from HRTEM exhibited a rod-like structure for ZnO with uniform length and width. The synthesized ZnO nanorods exhibited a bandgap of 2.68 eV, and hence it can be used for various potential applications.

Acknowledgements The authors would like to acknowledge the PG Department of Physics, Basaveshwar Science College, Bagalkot, Karnataka, India, for providing the lab facility to carry out this work.

References

1. Yi GC, Wang W, Il P (2005) ZnO nanorods: Synthesis, characterization and applications. Semicond Sci Techno 20:S22. https://doi.org/10.1088/0268-1242/20/4/003
2. Aspoukeh PK, Barzinjy AA, Hamad SM (2021) Synthesis, properties and uses of ZnO nanorods: a mini review. Int Nano Lett 2:153–168. https://doi.org/10.1007/s40089-021-00349-7
3. Fried JP, Swett JL, Bian X, Mol JA (2018) Challenges in fabricating graphene nanodevices for electronic DNA sequencing. MRS Comm 8(3):703–711. https://doi.org/10.1557/mrc.2018.187
4. Wendt JB, Ieee SM, Potkonjak M, Mcdonald N, Rose GS, Wysocki B (2015) Nano meets security: exploring nanoelectronic devices for security applications. IEEE Proc 103(5):829–849. https://doi.org/10.1109/JPROC.2014.2387353
5. Klimovskaya AI, Ostrovskii IP, Ostrovskaya AS (1996) Influence of growth conditions on morphology, composition, and electrical properties of n-Si wires. Phys Status Solidi A 153:465–472. https://doi.org/10.1002/pssa.2211530221
6. Zhang X, Wu D, Geng H (2017) Heterojunctions based on II–VI compound semiconductor one-dimensional nanostructures and their optoelectronic applications. Crystals 7(10):307. https://doi.org/10.3390/cryst7100307
7. Huang H, Kerr T (2002) Nanostructured composites: a high capacity, fast rate $Li_3V_2(PO_4)_3$/carbon cathode for rechargeable lithium batteries. Adv Mater 14(21):1525–1528. https://doi.org/10.1002/chin.200303011
8. Labis JP, Al-Anazi AQ, Al-Brithen HA, Hezam M, Al-Garni A, Al-Harbi AA, Al-awadi AS, Khan A, El-Toni AM (2019) Designing ZnO nanostructures (nanoworms, nanoflowers, nanowalls, and nanorods) by pulsed laser ablation technique for gas-sensing application. J Am Ceram Soc 102:4367–4375. https://doi.org/10.1111/jace.16270
9. Lepot N, Van Bael MK, Van Den Rul H, Haen JD (2007) Synthesis of ZnO nanorods from aqueous solution. Mater Lett 61:2624–2627. https://doi.org/10.1016/j.matlet.2006.10.025
10. Anastasiou A, Lee MHJ, Leach C, Freer R (2004) Ceramic varistors based on ZnO–SnO_2. J Eur Ceram Soc 24:1171–1175. https://doi.org/10.1016/S0955-2219(03)00472-2
11. Jiaqiang X, Yuping C, Daoyong C, Jianian S (2006) Hydrothermal synthesis and gas sensing characters of ZnO nanorods. Sens Actuators B: Chem 113:526–531. https://doi.org/10.1016/j.snb.2005.03.097
12. Yoshihara J, Campbell CT (2021) Copper nanoparticles with high index facets on basal and vicinal ZnO surfaces. J Phy Chem (ACS) 42:23561–23569. https://doi.org/10.1021/acs.jpcc.1c08008
13. Kanavi PS, Meti S, Fattepur RH, Patil VB (2022) Open nano emphasized temperature dependent electrical properties study of fabricated ZnO/PVA/PANI nanocomposite films. OpenNano 7:100057. https://doi.org/10.1016/j.onano.2022.100057

14. Meti UBKS (2020) Graphene-based ZnO nanocomposites for supercapacitor applications, graphene as energy storage materials for supercapacitors. 64:181–208. https://doi.org/10.21741/9781644900550-7
15. Li XD, Chen TP, Liu P, Liu Y, Leong KC (2013) Effects of free electrons and quantum confinement in ultrathin ZnO films: a comparison between undoped and Al-doped ZnO. Opt Express 21:4000–4003. https://doi.org/10.1364/OE.21.014131
16. Mydeen SS, Kumar RR, Kottaisamy M, Vasantha VS (2020) Biosynthesis of ZnO nanoparticles through extract from Prosopisjuliflora plant leaf: antibacterial activities and a new approach by rust-induced photocatalysis. J Saudi Chem Soc 24:393–406. https://doi.org/10.1016/j.jscs.2020.03.003
17. Vivek NSE, Meena MSM, Vetha MVI (2017) Synthesis of ZnO nanorods by one step microwave-assisted hydrothermal route for electronic device applications. J Mater Sci Mater Electron 29:2927–2938. https://doi.org/10.1007/s10854-017-8223-5
18. Altintas O, Liu Y, Petford-Long AK (2015) Synthesis of uniformly distributed single- and double-sided zinc oxide (ZnO) nanocombs. J Cryst Growth 430:34–40. https://doi.org/10.1016/j.jcrysgro.2015.08.007
19. Duan X, Huang Y, Cui Y, Wang J (2001) Indium phosphide nanowires as building blocks for nanoscale electronic and optoelectronic devices. Nature 409:66–69. https://doi.org/10.1038/35051047
20. Dash P, Manna A, Mishra NC, Varma S (2018) Synthesis and characterization of aligned ZnO nanorods for visible light photocatalysis. Phys E Low-Dimens Syst Nanostruct 107. https://doi.org/10.1016/j.physe.2018.11.007
21. Bhat TS, Bhogale SB, Patil SS, Pisal SH, Phaltane SA, Patil PS (2020) Synthesis and characterization of hexagonal zinc oxide nanorods for Eosin-Y dye sensitized solar cell. Mater Today Proc. https://doi.org/10.1016/j.matpr.2020.08.687
22. Hassan N, Shahat A, El-Didamony A, El-Desouky MG, El-Bindary AA (2020) Synthesis and characterization of ZnO nanoparticles via zeoliticimidazolate framework-8 and its application for removal of dyes. J Mol Struct 1210:128029. https://doi.org/10.1016/j.molstruc.2020.128029
23. Lu H, Deng K, Shi Z, Liu Q, Zhu G, Fan H, Li L (2014) Novel ZnO microflowers on nanorod arrays: local dissolution-driven growth and enhanced light harvesting in dye-sensitized solar cells. Nanoscale Res Lett 9:1–8. https://doi.org/10.1186/1556-276X-9-183
24. Jayarambabu N, Rao KV, Prabhu YT (2015) Beneficial role of zinc oxide nanoparticles on green crop production. Int J Multi Adv Res Trends 273–282. ISSN 2349-7408
25. Rathidevi K, Velmani N, Tamilselvi D (2019) Electrical conductivity study of poly (p-anisidine) doped and undoped ZnO nanocomposite. Mediterr J Chem 9:403–410. https://doi.org/10.13171/mjc01912071050kr
26. Kalpana VN, Patra RR, Rajeswari VD (2017) Green synthesis of zinc oxide nanoparticles using fresh stem of cissus quadrangularis extract and its various in vitro studies. Asian J Chem 29:1323–1327. https://doi.org/10.14233/ajchem.2017.20483
27. Fooladsaz K, Sayad A, Jamaleddini A (2012) Dopamine determination with a biosensor based on catalase and modified carbon paste electrode with zinc oxide nanoparticles. Int J Electrochem Sci 7:9892–9908. https://doi.org/10.1016/S1452-3981(23)16246-8

Eco-friendly Chemical Polishing of Copper and Copper Alloys

Meera Srinivas

1 Introduction

Copper and its alloys are widely used in electronics, architecture, plumbing, transportation, renewable energy, medical equipment, coins, etc. [1]. Different types of copper alloys include brass with Zn as the alloying element, phosphor bronze with tin as the principal alloying element, cupronickel with nickel as the principal element, silicon bronze containing Si as the principal alloying element and so on [1]. However, copper metal and its alloys are slowly oxidized when they are exposed to the atmosphere for long periods of time forming a blue-green coating referred to as patina [2]. In Cu-based alloys a redox electrochemical reaction takes place due to the interaction between the alloys and the environment. Based on the nobility of the metal or the alloy, the chemical–physical properties of the material get altered. The artefacts may undergo uniform or localized corrosion. Corrosion due to the atmosphere results in uniform corrosion, this may be aggravated due to the presence of sulphur oxide, powder particles and hygroscopic materials. This results in the change in the copper alloy products (Cu_2O, CuO, Cu_2S, CuS, $Cu_4(SO_4)(OH)_6$, $Cu_3(SO_4)(OH)_4$, etc.) formation depending on the temperature, pH, O_2 content, amount of salt, etc. [3]. Over long periods, noble and non-noble patinas are formed on the copper artefacts. Similarly, some of the copper and copper alloys used in industries also undergo corrosion. It is detrimental to preserve artefacts and pass them on to future generations. Therefore, they are subjected to a cleaning process using chemical or mechanical methods [4, 5].

The above issue can be overcome by using different types of polishing as depicted in Fig. 1 [6]. Mechanical polishing involves cutting and plastic deformation of the surface of the material using abrasives, felt wheels and emery papers manually or using auxiliary tools. This process enables a surface roughness of 0.008 μm. On the

M. Srinivas (✉)
M/s. Pyro Technologies, Kanakapura, Karnataka, India
e-mail: meera@pyrotechnologies.co.in

S. T. Aruna et al. (eds.), *Recent Advances in Electrochemical Science and Technology*, Springer Proceedings in Materials 47, https://doi.org/10.1007/978-981-97-3958-5_15

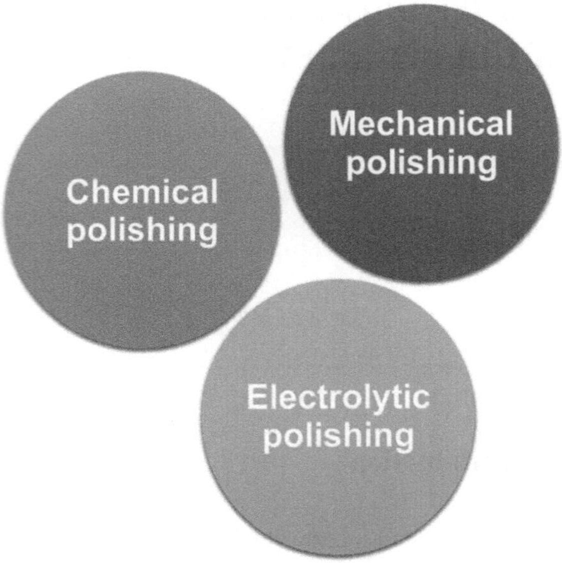

Fig. 1 Schematic depicting various types of polishing

other hand, in chemical polishing as the name indicates, a chemical medium is used to dissolve the protruding material surfaces thereby resulting in a smooth surface finish and deburring metal surfaces. This process results in the formation of a passivation layer due to the effect of solutions and galvanic couples. This method is very fast and surfaces can be cleaned immediately. Electrolytic polishing, also referred to as anodic polishing is similar to chemical polishing. However, the cathode reaction is not active in this process and results in a better-polished surface. The advantages and disadvantages of different polishing methods are summarized in Table 1 [6–8].

In general, surfaces of copper and its alloys before electroplating were pre-treated by dipping in bright baths containing one or more acids such as H_2SO_4, HCl, HNO_3 and chromic acid. However, these baths are harmful, toxic and are polluting nature as disposal of spent solutions is always a major concern. The baths based on HNO_3, H_3PO_4 and CH_3COOH require high operating temperatures in the range of 50–80 °C and vigorous mechanical stirring and results in emissions of toxic gases. With increased environmental regulations, the concept of green or eco-friendly is in vogue and researchers are working towards materials and methods that are harmless to the environment and the person handling them [9]. Apart from this, many of the cultural heritages have to be protected and preserved from the environment [10, 11]. In chemical polishing, chemical compounds that can complex easily with copper ions are selected. In recent work, amino acids have been explored for cleaning bronze artefacts to achieve maximum chelation [12]. Usually, the chemical cleaning process employs chelating solutions containing ethylenediaminetetraacetic acid (EDTA). Though EDTA is not harmful to humans, it is not biodegradable and is resistant to water purification treatments as it gets involved in the mobility of heavy and radioactive ions resulting in water eutrophication [13]. In some of the chemical polishing

Table 1 Advantages and disadvantages of mechanical, chemical and electrolytic polishing methods

Mechanical polishing		Chemical polishing		Electrolytic polishing	
Advantages	Disadvantages	Advantages	Disadvantages	Advantages	Disadvantages
Good brightness	Labour intense	Efficiency is higher and complex shapes can be polished	Cannot be heated	Linger lustre time	High capital investment and complex pre-polishing
Good surface finish	Complicated parts cannot be polished	Multicomponent polishing at a time	Harmful gas may be released	Consistent colours throughout	No universal electrolyte
Good surface cleanliness	Can undergo corrosion	No capital investment	Difficult to regenerate polishing solutions	Less pollution	For complex parts tooling and electrodes are required
Aesthetically appealing	Mechanical strength reduced	Improved corrosion resistance	Inhomogeneous brightness	Improved corrosion resistance	Cooling facilities required for industrial-scale operation

solutions, the artefacts are immersed in oxidizing solution or the solution is sprayed onto the surfaces. The oxidizing solutions are peroxide-based and they contain other additives such as stabilizers, brighteners, inhibitors, surfactants, etc. [14]. There are several patents on compositions of baths for chemical polishing [15–17].

Currently, eco-friendly baths containing organic acids, triggered with H_2O_2 (hydrogen peroxide) and other additives are becoming very popular. This paper reports some efforts made in this direction by using environmentally friendly chemicals. The proprietary bath has been upscaled and commercialized and is being used in many industries. The advantages of the eco-friendly polishing solution are as follows: (i) all the raw materials are locally available and are cost-effective, (ii) it has a good shelf-life, (iii) easy to handle and easy to store, (iv) safe to the environment and easy to dispose of the solution, (v) the product is stable and (vi) the quality of the treated parts is excellent. More details regarding the chemical bath and some treated surfaces and their finish are presented in this paper. Also, an attempt is made to present the effect of different constituents of the chemical bath on the final finish/polishing of the artefacts.

2 Experimental

In the chemical polishing process, a chemical reaction takes place between the material (artefact) and the applied etchant which results in the removal of loose powder particles. The effectiveness of the process depends on the nature of the etchant, the temperature of the etching, the applied agitation and the duration [18]. In the present

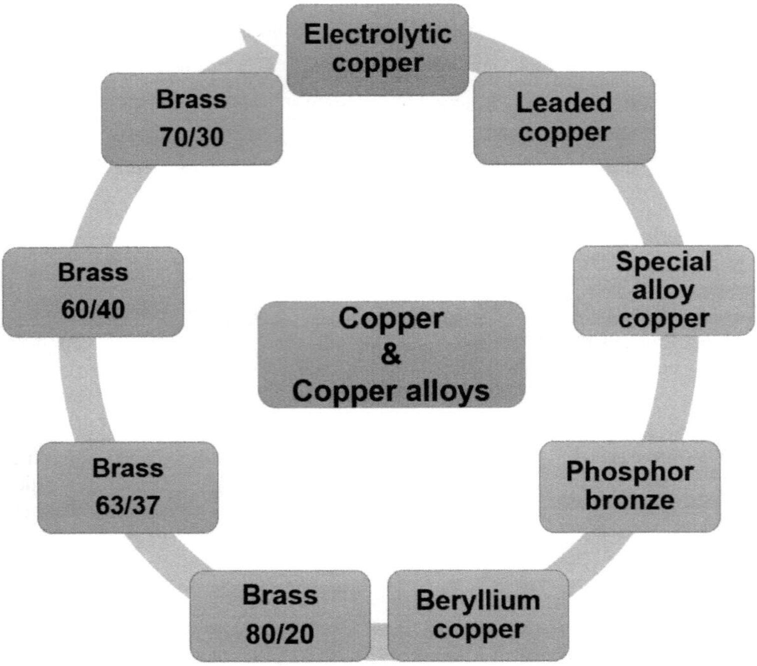

Fig. 2 Schematic representing various copper and copper alloys that were subjected to successful chemical polishing

study, chemical polishing was carried out on copper and various copper alloys as shown in Fig. 2. Eco-friendly chemical polishing of copper and copper alloys is a two-step process using a proprietary bath developed in the industry.

The polishing bath, consisting of primarily an organic acid or mixture of organic acids, an oxidizing agent, a wetting agent, inhibitors, stabilizers, brightening additives and demineralized water, was developed. All the ingredients used are environmentally friendly. Table 2 summarizes the constituents of the chemical polishing bath along with their concentration. The bath effectively works at a pH value of 1–4, together with an oxidizing agent and one or more stabilizers, brighteners and other additives. The source of acidity is selected from organic acids like citric acid, DL-malic acid, tartaric acid, acetic acid, lactic acid, glycolic acid and so on (Fig. 3). Organic acids are preferred as etchants as they act as chelating agents and the metal removal is very slow and thereby ensures uniform polishing. The above-mentioned various organic acids alone or in combinations with one or more organic acids were used for the experiments. The oxidizing agent is hydrogen peroxide (H_2O_2) of 35% concentration and stabilizers are selected from polymers, alcohols, organic acids, triazoles and other proprietary additives. Stabilizers reliably suppress the decomposition of the peroxide. Besides, the selection of the stabilizers is crucial as a wrong stabilizer selection can lead to a violent bath reaction and sometimes boiling and overflowing of the solution. This is definitely a safety concern and hence care should

Table 2 Constituents of the chemical polishing bath

Sl. no	Ingredients	g/L
1	Organic acids	50–300
2	35% hydrogen peroxide	80–150
3	Wetting agents	5–20
4	Solvents	2–10
5	Brightening additives	10–100
6	Inhibitors	1–5
7	Stabilizers	50–150
8	Demineralized water to make up to 1 L	Balance

be taken while choosing the stabilizer. Wetting agents are selected from biodegradable non-ionic surfactants like coco-glucoside, decyl glucoside, lauryl glucoside and so on. These are surfactants that are generally used in baby products, bath products, cleansing products, skin care products, eye makeup and hair care products. Corrosion inhibitors like 1, 2, 4-triazole (TAZ) are used in the bath apart from proprietary stabilizers and brighteners. Glycols are in general used as brighteners. A known stabilizer for peroxides is urea, which, according to the state of the art, is added both on its own and in conjunction with other stabilizers, e.g. ammonium ions, to aqueous peroxide-containing solutions.

It is important to know that many processes in which aqueous peroxide-containing solutions are used in chemical polishing are exothermic. Therefore, there are chances of the solutions getting heated up during the process and must therefore be cooled in order to maintain the desired operating temperature. For aqueous hydrogen peroxide solutions, this critical temperature is around 65 °C. Irreversibly, and the solutions/polishing baths have to be discarded, disposed of and fresh solutions must be

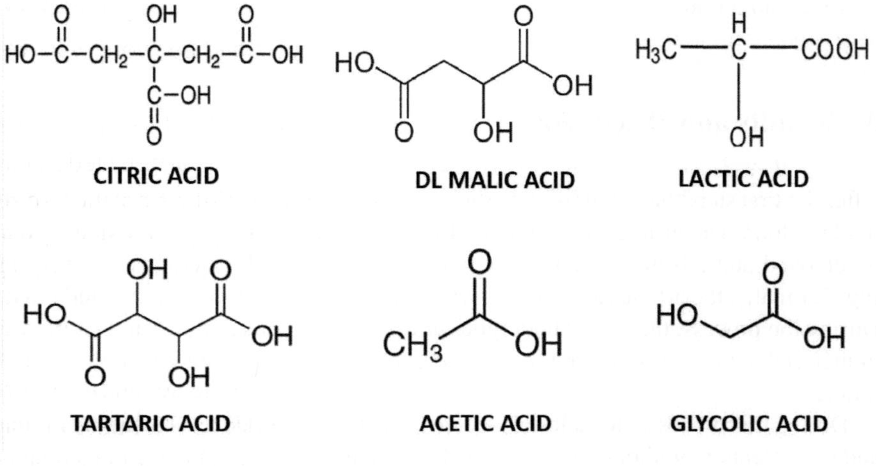

Fig. 3 Chemical structures of various organic acids used in chemical polishing

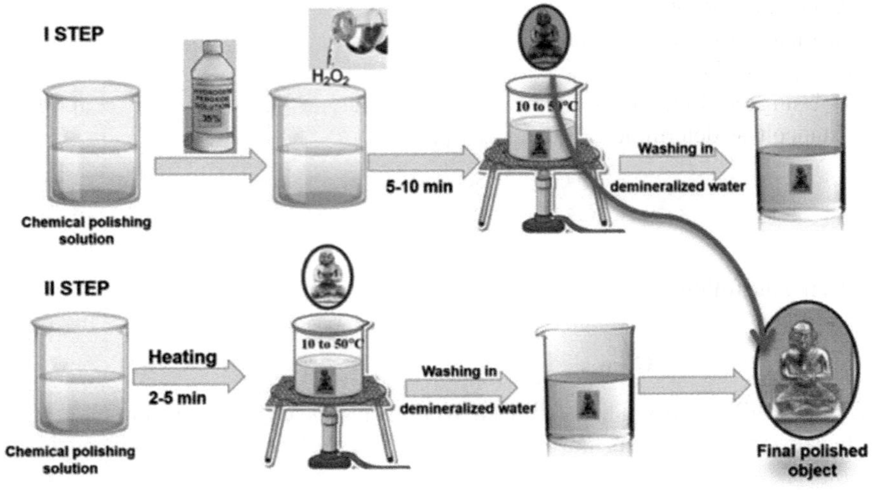

Fig. 4 Schematic depicting the steps involved in the chemical polishing process

prepared. All the other constituents used in the chemical bath are proprietary in nature and hence the details are not disclosed in Table 2.

The steps involved in chemical polishing are illustrated in Fig. 4. The chemical polishing process involves two steps: In step 1, the eco-friendly chemical polishing solution is activated with hydrogen peroxide. The object is dipped in solution or the solution is sprayed onto the artefact by maintaining the temperature of the solution in the range of 10–50 °C for 5–10 min. After this, the article is removed from the bath and rinsed thoroughly with demineralized water. In Step 2, the eco-friendly chemical polishing solution without hydrogen peroxide is used for immersing the artefact in a tank, or the solution is sprayed. The temperature is maintained between 10 and 50 °C for 2 and 5 min.

3 Results and Discussion

After the first step, the colour of the articles turns dark brown or similar to the bronze oxide colour. The articles are removed from the bath and the bronze oxide colour is removed and a brilliant high gloss polished surface is obtained as evident from Fig. 5. Finally, the articles are rinsed thoroughly with demineralized water and dried. During the process, the article undergoes uniform oxidation resulting in the removal of material in the nanoscale and forming a bronze colour on the surface of the treated article.

During this process, the material is removed at the nanoscale level resulting in spic and span cleaning without affecting any dimensional changes. The rate of etching is 0.30 μm/min at 40 °C. The etching rate in nitric acid-based solution is ~3.5 μm/min

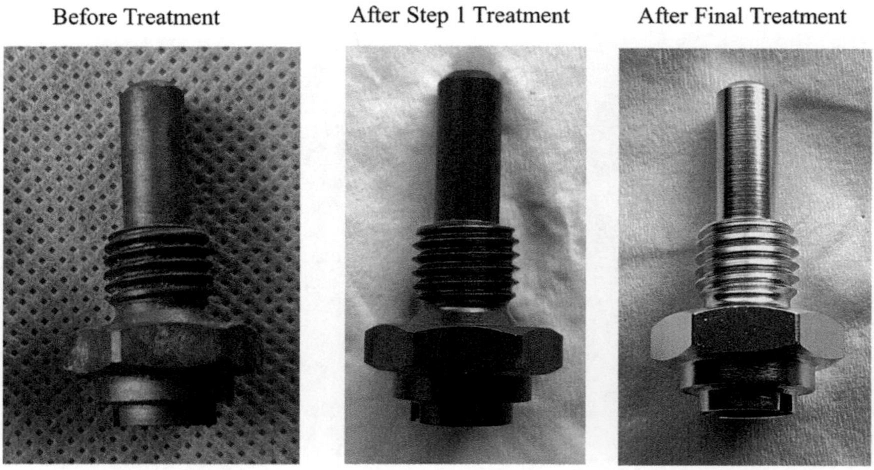

Fig. 5 Photograph of a copper component **a** before and **b** after chemical polishing

at 25 °C for 3 min, whereas in the eco-friendly polishing solution, the etch rate is ~0.32 μm/min at 25 °C for 3 min. In a few cases, the surface roughness value (Ra) has reduced from 167–195 to 60–75 μm resulting in an ultra-high glossy mirror-like finish. The process gives superior mirror-like glossiness which is unmatched by any other conventional chemical brightening/polishing process. It should be noted that with 1 L of electropolishing solution, 150 sq. ft of the area can be polished. Figure 5 shows the photographs of a component subjected to chemical polishing. Similarly, Figs. 6 and 7 depict the photographs of some industrial components and temple artefacts before and after chemical polishing, respectively (Fig. 8).

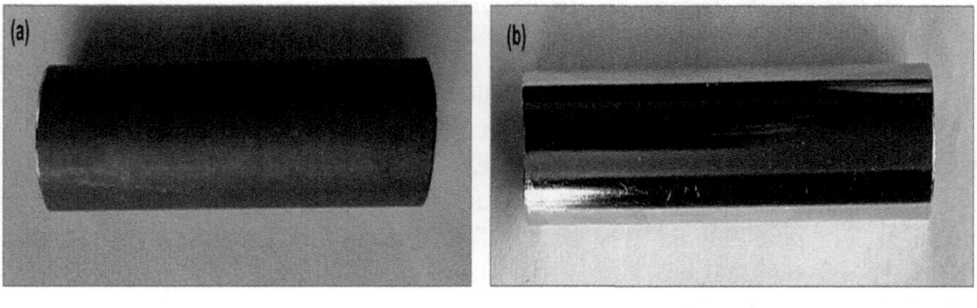

Fig. 6 Photograph of a copper component **a** before and **b** after chemical polishing

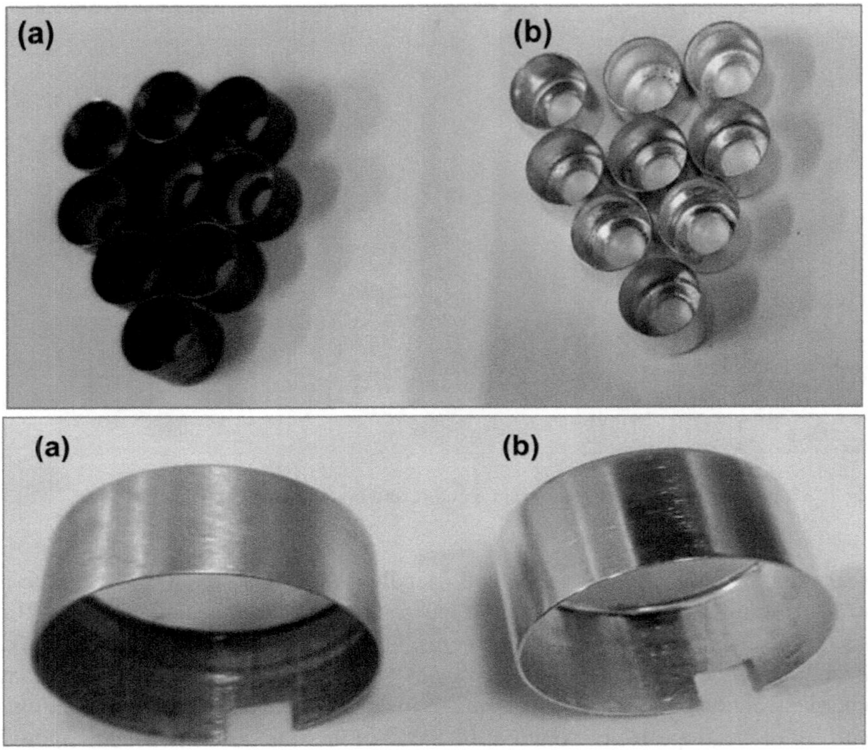

Fig. 7 Photographs depicting various brass items **a** before and **b** after chemical polishing

3.1 The Effect of % of Various Constituents of the Chemical Polishing Solution on the Product Finish

In general, it was observed that the constituents of the bath and their % influenced the nature of polishing. The effect of different % of the constituents of the bath on the nature of polishing is summarized in Table 3. From the table, it may be generalized that a lower % of ingredients leads to matt finishing, a medium % of ingredients results in good polishing and a higher % of ingredients, in particular, DL malic acid, hydrogen peroxide and proprietary stabilizers and brightening additives results in high gloss mirror finishing.

3.2 Bath Maintenance and Disposal

Some of the reported baths cannot be reused for polishing surfaces and replenishment of the bath is difficult due to severe changes in the local compositions of the bath. The

Fig. 8 Photographs depicting various temple artefacts **a** before and **b** after chemical polishing

Table 3 The effect of the various constituents of the chemical polishing solution and its % on the nature of polishing

Sl. No	Ingredients % by weight	Matt finishing (%)	Good polishing (%)	High gloss mirror polishing (%)
1	DL malic acid	8	14	20
2	35% hydrogen peroxide	8	10	12
3	Ethyl alcohol	2	4	6
4	Propylene glycol	2	4	4
5	Coco-Glucoside	1	2	2
6	1,2,3-Benzotriazole	1	0.50	0.50
7	Proprietary stabilizers, and brightening additives	4	6	8

bath reported in this paper is simple and easy to maintain. Due to the depletion of the peroxide concentration in the solution due to catalytic decomposition, the process quality will be adversely affected. Therefore, hydrogen peroxide has to be added before reusing the bath. The same chemical bath solution can be used at least thrice by adding hydrogen peroxide solution to it. As soon as this critical concentration of metal ions is reached in the aqueous peroxide-containing solution, as a rule, these solutions can no longer be used and must be discarded, often at considerable cost, both for the disposal of the old solutions and for the preparation of new, fresh Solutions. The critical upper limits for the content of metal ions depend on the type of metal ions, the composition of the peroxide-containing Solutions and the stabilizers used. The values are, for instance, in the range of 10–30 g/L for copper and 20–40 g/L for iron. The spent solution can be treated and discharged easily. The copper and other solid precipitates from the spent solution can be filtered and separated successfully on the lab scale. The pH of the filtrate is adjusted to pH 7 by using a suitable alkali and disposed of by following the practices adopted for disposal of solutions. The process and solution meet the global standards for QSHE (Quality, Safety, Health and Environment).

4 Conclusions

This study has demonstrated the application of an eco-friendly polishing solution for copper and copper alloys to clean, remove tarnish/corrosion and brighten the surfaces to their previous appearance. Chemical polishing has been established using a simple immersion process that works at ambient temperature. Highly cost-effective mirror finishes with Ra value 6–12 μm have been achieved in a few cases. The proprietary bath, process and solution are working successfully and are being widely used in industries. The developed process and solution can be successfully implemented for both aesthetic and functional requirements to treat/polish copper and copper alloys. The cleaning solutions described in literature are regularly used at temperatures in the range from 80 to 100 °C; whereas the current bath can be used at room temperature. The highlight of the developed solution is that the effluent treatment of the spent solution is simple and economical. Since the solutions used are eco-friendly, there is no negative impact on the environment and health of the people handling the solutions.

Acknowledgements The author expresses her deep gratitude to Mr. Rajeeva Deekshit, CMD M/s. Pyro Technologies, Kanakapura, for guiding her and making the project a successful one.

References

1. Joseph R, Davis (ed) (2001) Copper and copper alloys. ASM, Materials Park, OH, USA
2. Kuntzleman TS, Cullen DM, Scott M, Ragan D (2020) Rapid formation of copper patinas: a simple chemical demonstration of why the statue of liberty is green. J Chem Educ 97(8):2244–2248. https://doi.org/10.1021/acs.jchemed.0c00417
3. Brambilla L (2011) Multianalytical approach for the study of bronze and gided bronze artefacts [doctoral thesis]. Universita degli Studi di Milano, Milan, Italy
4. Petiti C, Toniolo L, Gulotta D, Mariani B, Goidanich S (2020) Effects of cleaning procedures on the long-term corrosion behavior of bronze artifacts of the cultural heritage in outdoor environment. Environ Sci Pollut Res Int 27(12):13081–13094. https://doi.org/10.1007/s11356-020-07814-4
5. Parisi EI, Bonelli N, Giorgi R, Ingo GM, Baglioni P (2017) Development of an innovative film-forming cleaning system for the removal of corrosion products from copper alloy artifacts
6. Bridgland J (ed) ICOM-CC 18th triennial conference preprints; Sep 4–8; Copenhagen, Denmark. https://www.cnclathing.com/guide/types-of-polishing-process-advantages-disadvantages-uses-working-principle-cnclathing
7. https://www.yhabrasives.com/News/Advantages-and-disadvantages-of-three-polishing-methods-for-stainless-steel.html
8. https://www.sansmachining.com/polishing-process-type-advantages-application-and-difference-between-grinding/
9. Kolodynska D (2011) Chelating agents of a new generation as an alternative to conventional chelators for heavy metal ions removal from different waste waters. Expanding issues in desalination. InTech Publishers, pp 339–371
10. Macchia A, Sammartino MP, Laurenzi Tabasso M (2011) A new method to remove copper corrosion stains from stone surfaces. J Archaeol Sci 38(6):1300–1307. https://doi.org/10.1016/j.jas.2011.01.005
11. Mezzi A, Angelini E, De Caro T, Grassini S, Faraldi F, Riccucci C, Ingo GM (2012) Investigation of the benzotriazole inhibition mechanism of bronze disease. Surf Interf Anal 44(8):968–971. https://doi.org/10.1002/sia.4841
12. Macchia A, Colasanti IA, Rivaroli L, Favero G, de Caro T, Munoz LP, Campanella L, Russa MFL (2023) Natural based products for cleaning copper and copper alloys artefacts. Nat Prod Res 37(7):1177–1184. https://doi.org/10.1080/14786419.2021.2000408
13. Masoudi F, Shirvani M, Shariatmadari H, Sabzalian MR (2020) Performance of new biodegradable chelants in enhancing phytoextraction of heavy metals from a contaminated calcareous soil. J Environ Health Sci Eng 18(2):655–664. https://doi.org/10.1007/s40201-020-00491-y
14. Stabilizer for acidic, metal-containing polishing bath, US 2008/0067.149 A1 (2007)
15. Solutions for chemically polishing surfaces of copper and its alloys US4086176A (1975)

16. Baths and process for chemical polishing of copper or copper alloy surfaces United States Patent 5098517 (1990)
17. Method for chemically polishing copper or copper alloy, US3556883A (1967)
18. Chmielewska A, Jahadakbar A, Wysocki B, Elahinia M, Święszkowski W, Dean D (2022) Chemical polishing of additively manufactured, porous, nickel–titanium skeletal fixation plates. 3D Print Addit Manuf 9:269–277. https://doi.org/10.1089/3dp.2020.0209

GPSR Compliance

The European Union's (EU) General Product Safety Regulation (GPSR) is a set of rules that requires consumer products to be safe and our obligations to ensure this.

If you have any concerns about our products, you can contact us on ProductSafety@springernature.com

In case Publisher is established outside the EU, the EU authorized representative is:

Springer Nature Customer Service Center GmbH
Europaplatz 3
69115 Heidelberg, Germany

Batch number: 10703869

Printed by Printforce, the Netherlands